An Avonridge Novel

M.B. NASDEO

The Reading Glass Books
1-888-420-3050
www.readingglassbooks.com
fulfillment@readingglassbooks.com

Dedication

This book is for Alice, the true Queen in my life.
Your gentle strength is my guiding force. Thanks Mom

ONE

Marissa walked into the kitchen and dropped her keys onto the table. Merlin sat on the counter and she reached out and absentmindedly scratched him behind the ear, causing him to erupt in motor-boat-like purrs. Her day had been just like yesterday, and the day before, and probably the way the day would go tomorrow. Nothing seemed to change in Marissa's world. The only bright spot was the upcoming Renaissance Faire. There was always magic at the Renaissance Faire, and she would lose herself completely within the realm of kings and queens and knights and magicians. She looked forward to that with a passion.

Opening the refrigerator door, she pondered the toughest question she'd had to face all day; what to eat for dinner. The fridge disappointed her again, as it usually did. Not being one to fix elaborate meals for herself, there were rarely any leftovers. She generally liked leftovers, it was so easy just to throw something in the microwave.

She heaved a melancholy sigh and shut the refrigerator door, then ambled into the living room to sit in front of the television, Merlin trailing along behind. As she sat down on the couch and grabbed the remote, Merlin jumped onto her lap, purring heartily and rubbing his head against her shoulder. She stroked his soft, black and white fur, and clicked on the TV. The screen erupted, filling the room with an eerie glow, and Marissa settled in for an evening of mind-numbing relaxation. However, even reruns of old sitcoms she had grown to memorize and enjoy, couldn't hold her attention.

Marissa's mind kept replaying the ordinary, common events of her day. She worked for the local telephone company in the internet maintenance department, answering calls from customers who were having trouble with their internet. Every day it was the same; complaints, complaints, complaints. She was tired of hearing people bitch and complain all day long, but there was very little movement within the ranks of the company at the moment, and she was stuck. Her only consolation was the fact that the pay was pretty good for not having gone to college. She worked hard, but the company provided a decent living. Unfortunately, the decent living didn't make up for the fact that she had no life.

She didn't know exactly where her train had been derailed, but Marissa was fast approaching the time when she would look back on her life and wonder where it had gone. She'd never married, even though she had wanted a family. Oh, there had been chances, but somehow, she never thought the men were exactly what she wanted. At the time she thought she was being selective, but now, looking back, she wondered if she was too judgmental, too demanding. Had she expected too much? She only wanted someone she could laugh with, enjoy life with. Her sense of humor was full and rich, and she needed someone with the same.

She had thought she'd found him in Eric. Ah, yes, Eric. Tall, had to be six foot three, and the classic textbook good looks; black hair, warm, sky-blue eyes, skin smooth and fair. Tall, dark and handsome. And he had loved Marissa, at least he had said so. She'd never been very popular, having gone through her whole high school career without a boyfriend, and only having a few meaningful relationships since. When Eric first caught her eye, she was standing on line at the convenience store checkout, thumbing through a tabloid. He stepped up behind her with a cup of coffee in his hand and smelling of musk, her favorite cologne. She looked up at him and smiled. He looked her directly in the eye and smiled back, causing her heart to skip a beat. Marissa always liked it when a man looked her in the eye, it made him appear honest, straightforward. If there was one thing she hated in anyone, male or female, it was insincerity. If you had a problem with her, tell her to her face, don't hide your feelings, Marissa never hid hers.

And, in not hiding her feelings, she decided to give Eric a try.

"I've never seen you here before," she said. "Off your usual route, or just running late?"

"I just switched jobs, and this is my first day going this way," he replied, his voice just deep enough to be sexy without arrogance. "Do you stop here every morning?"

"Usually, if I'm not running late," she laughed.

Marissa didn't rush things. She met Eric almost every morning from then on, gradually getting to know him. A few weeks later they met for lunch. A week after that they had a romantic dinner. It didn't take very long for the friendship to blossom into a relationship.

Right from the start it seemed as if they had known each other forever. They laughed at the same things, saw life in the same light. They could even finish each other's sentences almost from the beginning. Marissa was in Heaven, having finally found someone she wanted to spend the rest of her life with. It seemed too good to be true. Unfortunately, when it seems so, it usually is; too good to be true, that is. And this was not the exception.

She and Eric moved in together, renting a lovely little garden apartment about halfway between their two jobs. At first everything was perfect. She slept on the left side of the bed; he preferred the right. He liked her cat Merlin, she loved his Corvette, Rose. She found it cute that he named his car, but then she found almost everything about Eric cute; from the way his lips would move when he read the Wall Street Journal, to the way he fell asleep watching the late-night news in bed. The word idyllic often came to mind when Marissa thought about her relationship with Eric. At least in the beginning. Eventually, however, idyllic was replaced with words much less appealing, words such as dishonest, duplicitous, deceitful.

Sitting in her living room with Merlin on her lap and the T.V. playing an old Happy Days rerun, Marissa couldn't even remember the way she had found out. The way she found out was really not important, considering the what of what she found out. Somehow, some way, she found out Eric had been in prison. Now, having committed a grievous mistake, and having paid your dues for said grievous mistake was not reason enough to write someone off in Marissa's mind. Everyone made mistakes, some just were unlucky enough to make devastating ones. The fact that he had not trusted her enough with the information to give it freely, but tried to hide it from her, was what mattered to Marissa. Eric had not had the decency to tell her himself of his past, but she found out through some other means. She couldn't remember if a friend of his had spilled the beans, or he blurted it out himself, but she could remember the shock of finding out that he didn't trust her.

Oh, he said he was afraid of how she would take it, he said he would have told her eventually, when he knew she could handle it, it's just that it wasn't the right time. Marissa asked when would be the right time to find out he had killed someone? That's something you have to make the time for, it doesn't just happen.

Barroom fights like the one that ended a man's life just happen. That kind of thing is spur of the moment. Men drink, get loud, get stupid and end up sending another drunk, stupid man through a plate glass door. Grievous mistake, but mistake nonetheless. No, the 'right time' for telling one's lover about that barroom fight doesn't just happen. One must make the time to offer that information.

At first Marissa was unsure of what to do. Was Eric an alcoholic that would make that same mistake again? They had been together for over three years when she found out this little tidbit. During that time they had been to parties, gone to bars, been any number of places where Eric drank. She had never seen him lose control of himself, had never seen him become sloppy or stupid. With a lot of soul-searching, she could chalk the incident up to a one-time thing. But they had been together for over three years, and he had not offered up this information willingly. Marissa had found out by mistake, and that troubled her more than the incident itself.

The rift caused by this newfound knowledge slowly tore the relationship apart. Marissa and Eric no longer seemed to see life in the same light. They stopped knowing what the other was thinking, were no longer able to finish each other's sentences. Where once had been a communion of two hearts, a meeting of two souls, now there was a void. One day Eric quietly packed and walked out the door. Marissa couldn't even remember if he had said good-bye. It just happened like the ticking of a clock, as if it simply was time for him to go. And he was gone from her life forever.

Since then, Marissa had had a few lovers, but none shared her home. She had become just a little more suspicious of everyone, just a little less trusting, more cynical. She hated the word 'jaded', but if she was to be honest with herself, jaded she had become.

Now, with her fortieth birthday just a few years away, she knew in her heart, even if she refused to admit it to her head, that she would probably never marry, never have children, would, in a few years, be the crazy Old Lady Brewer living with eighty-seven cats and throwing rocks at the kids that stole her roses. She supposed that wouldn't be so bad, but if she expected to have eighty-seven cats, she ought to start bringing some home. Merlin couldn't be called upon to fulfill that feat all on his own.

Marissa finally gave in and turned the T.V. off. She hadn't realized she'd been sitting there for so long, and it was time to turn in. She hadn't even eaten dinner, but was feeling so bummed out, she didn't even care. Oh well, the Renaissance Faire was this weekend, and she would have fun at that. She had her medieval gown and cloak, her headpiece with silver and black cloth roses, and a great deal of anticipation. It didn't even matter that she would be going alone, once she stepped through the gate into medieval England, she would become a player in the show. Marissa would be a lady, walking among the knights, watching the jousting, marveling at the swordplay, laughing at the jesters. Yes, it would definitely lift her spirits, at least for the weekend.

TWO

"But why can't they come fix my internet today? I haven't been able to use it all day."

"I'm sorry, ma'am, but it's almost five o'clock and the technicians are finishing up. I can have someone there first thing in the morning to fix your internet line."

"That's not good enough," the aggravated voice insisted. "My computer's been down all day, I don't want to wait 'til tomorrow."

"Again, I apologize, ma'am, but this is the first time you've called. We didn't know your internet's been down all day. The next appointment I have is first thing in the morning," Marissa explained.

"I want my internet fixed tonight!" the woman continued. "I shouldn't have to go without it all night. I happen to be a very, very important person."

"Yes ma'am," Marissa answered calmly, "all our customers are very, very important."

And so her day had been again, just like yesterday, and the day before. Marissa was beginning to wonder if it was worth all the aggravation. What was the point to it all? Her life had become a never-changing series of complaints. After a while, it seemed even the voices were identical. She had begun to dread coming to work in the morning, knowing her whole day would be filled with calls from unhappy people, all demanding to be treated better than the next person.

Yes, she was sure that woman believed she was "very, very important." But, just as Marissa had answered, she tried to treat each caller as just as important. After all, she still remembered that the customers were the ones who actually paid her salary. Even so, it was difficult and disheartening to hear nothing but complaints all day long. She wished, yet again, that she could afford to quit her job and move somewhere where life was challenging and exciting. But, just like in *The Wizard of Oz*, Marissa believed that place was only in one's imagination.

At the end of the day, she took her handbag out of the drawer and headed for the door. Before she could make her escape, however, Rob intercepted her. He was the office letch, and he was constantly bragging about his conquests. It seemed he had bedded a different woman every night. He hit on Marissa every so often, but recently it seemed she had escaped his notice. However, now it looked as if her luck had run out.

"Hey gorgeous, how you doin'?" he said, and Marissa could almost see saliva dripping down his chin.

"Fine, Rob. Good night," and she tried to hurry past him. Unfortunately, he sidestepped and blocked her way.

"What's the hurry, beautiful?" he leered. "It's Friday night. Why don't me and you do something?"

"I've got plans, Rob. Please excuse me," and again she tried to get past. This time he put his arm across the doorway, leaning on one side of the jamb, and anchoring his palm against the other side.

"Break those plans, and have a good time for a change."

"With you?" she asked, looking him straight in the eye, her voice filled with sarcasm.

"Yeah, with me," he said. "Me and you could have a great time," and he licked his lips.

Marissa wanted to laugh in his face, wanted to tell him to go scratch, wanted to say any number of things to humiliate him and get him to leave her alone. Unfortunately, she had to work with him, and it was a tight-knit, if large, office. Mustering every ounce of self-control she could manage, she replied as benignly as possible.

"I can't break my plans, Rob. I have things I have to do that can't wait. Now please let me go home."

Looking only somewhat defeated, Rob dropped his arm and stood back.

"Don't make any plans for next Friday," he said. "It's me and you then."

"I really don't think so, I don't date men I work with," she lied, and hurried out the door. As she just about ran to her car, Marissa could hear Rob mutter, "stuck up," and she smiled to herself. After all, that was quite tame compared to what he could have called her.

Merlin sat on the counter as Marissa walked into the kitchen. She threw her keys on the table, and scratched the cat behind his ear. His appreciative purr rewarded her, and she hurried into the bedroom to try on her gown for the Renaissance Faire, so excited she felt almost giddy thinking about the days ahead. Tomorrow she would enter a world of knights and ladies, kings and queens, jesters and villains. Her heart beat fast in her chest, and her stomach felt like a cage of wild birds. She had toyed with the idea of bringing Merlin with her and, if he would walk on a leash, she might do just that.

She pulled the gown out of the closet and inspected it for any imperfections. Finding none, she pulled the headpiece off the shelf and gave it the once-over. Everything in fine order, everything ready for the big day.

Knowing she would find it difficult to get to sleep, Marissa drew a hot bath to try to relax, and settled into the lilac-scented bubbles. She let her imagination run free, conjuring up all the wondrous things she would find tomorrow at the Faire. She remembered from previous years the jousting, the sword fighting. Robin Hood and Maid Marion would be walking around, as would Friar Tuck, Little John, all the Merry Men.

Of course, the villains would be there as well. Marissa would encounter the Sheriff of Nottingham along the way. And one figure in particular stood out in her mind from last year's Faire. A tall knight, very tall, dressed all in black, with a black-leather mask covering his face completely. With knee-high leather boots and wicked looking sword strapped to his side, he looked every inch the evil nemesis of any law-abiding citizen. Just the thought of him made Marissa's heart beat faster. Something about him made her curious. He was hiding something, having covered his face so no one could recognize him absolutely, but there was something else. When she looked into his eyes that time last year, she could almost swear he was smiling at her.

The Black Knight's physique reminded her of Eric, which she found neither good nor bad. Eric was built exceptionally well. However, this knight was several inches taller than Eric. He had to be at least six and a half feet tall. And his eyes were dark, sapphire blue, not Eric's sky blue. Marissa chuckled as she remembered first spying the Black Knight. For a split second she was sure it was Eric, then she noticed the physical differences. It was a relief to realize it was not in fact Eric, because she could have fantasies about the knight. Had she believed him to be her old lover, she would have had to dismiss the knight from her mind, not wanting to dredge up old pains.

The mystery of the knight was intriguing. She knew people went to the Renaissance Faire for entertainment, to leave behind the everyday hassles and aggravations of real life and enter a world of fun and games. But there was something more to this knight. He had a presence, an aura that made him interesting. Of course, Marissa was sure that was his plan, to be alluring and intriguing. She was also sure he never wanted for female companionship when the sun went down and the Faire was over.

But Marissa was not so taken in by the Black Knight that she would even consider being that senseless. She lived in a world of diseases and assaults, and never let her guard down in such a stupid manner. Still, that's where fantasies come in. In fantasies, anything goes, and she let her fantasies run wild.

The sun rose in a clear sky, waking the morning birds and bringing life to the world. Marissa stretched and yawned. She was surprised to find that she had slept so well, and was now eager to start the day. Her dreams had been filled with visions of the knight on his black charger, sweeping her off her feet and taking her to a kingdom where she was treated like a royal lady. Marissa had always believed she had been born in the wrong century, and when she stepped through the gate into the Renaissance Faire, she felt truly at home. She had never revealed those feelings to anyone, but took joy in keeping them secret.

Tossing off the covers, she prepared to start her day. First thing was to get dressed for the Faire. Or perhaps, she ought to eat some breakfast. It was always a good idea to be fueled up when taking on the medieval duties of Merry Olde England. She ambled into the kitchen, Merlin rubbing up against her legs as she rummaged through the fridge. A fried egg sandwich would hit the spot, and she quickly fixed and ate her 'fuel'.

An hour after finishing her breakfast, Marissa stood before her full-length mirror, admiring the Renaissance Lady that stared back at her. She had wondered upon occasion if her demeanor might be too dark; the dress was black, long cape sleeves hanging gracefully to the floor; circular headpiece of black and gold cloth roses with black lace draped down her back. But she loved the looks she got from passersby when she strolled through the faire-grounds.

Her soft, brown hair fell half way to her waist, with gentle waves giving it an ethereal look. She arranged it around her shoulders, framing the hint of cleavage the gown allowed. She wore no makeup, preferring instead to let the excitement of the day add color to her cheeks. Light-brown eyes, bordering on hazel, looked back from the mirror, giving the appearance of gentle strength. Even though she stood only five feet five inches with a slender build, Marissa always liked the fact that she looked capable and strong. People rarely took advantage of her, somehow instinctively knowing she was no pushover.

Giving her image one last admiring look, she turned to Merlin, who was perched upon the bed, and said, "Shall we go, sir?"

He meowed in answer, and she clipped the leash to his collar, and together they headed out the door.

THREE

The day began cool and bright, with just a hint of a soft breeze that danced across Marissa's cheeks and lifted her hair around her headpiece. As she stepped through the gates of the Renaissance Faire, she drew in her breath in an excited gasp. It was just like walking through the looking glass into another realm, another time and place. If only time travel were possible, she would do anything in her power to go back to this time of lords and ladies, where chivalry was still alive and well. But she had mollified herself to visit, even if she couldn't actually live here.

Strolling along the dirt road, past the venders selling 'handmade' headpieces, Marissa and Merlin made a stunning pair; she as a lady dressed in black with her faithful, black-and-white cat beside her. Marissa marveled at the way Merlin seemed to sense the excitement of the day, strutting with his tail high in the air, as if he expected people to bow before him.

The little shops were mostly made of rough wood to look authentic, with rough wooden counters and support posts holding up wooden ceilings, and walled by elaborate cloth tarps. They were filled with treasures and trinkets of days of olde. There were handmade wooden shields with the crest of the lion, or an attacking dragon, all fashioned for the little ones to play knights in shining armor. Wooden swords hung on walls to accompany the shields, painted steel gray and black to imitate the genuine article.

Marissa stopped at a shop that sold chain-mail headdresses. She inspected one that framed a Styrofoam mannequin head, appreciating the detail of the workmanship. It seemed to be only tin, or some other soft, cheap metal, but it was exquisite. She wondered if she should consider buying it, but it really didn't go with her outfit. She was a lady, not a warrior. But still, perhaps next year she could change her appearance, go as the lady warrior, with leather armor, sword strapped to her side. Even though she was somewhat tiny for a role like that, it would be fun, and after all, the whole idea of the Renaissance Faire was to have fun. It would be something to think about.

While admiring the chain mail, Marissa noticed a woman asking the shopkeeper about a headdress. He answered her questions respectfully, but with very little attention.

As Marissa was about to leave the chain mail shop, the shopkeeper spotted her. He was a short, round, balding man, dressed as a smith with heavy, leather apron and large, calloused hands. His burlap pants were tied with a coarse rope knotted at the waist, and a once-white cotton shirt had been tucked in by clumsy hands.

"Milady," he said, his voice emitting obvious reverence. "It is good to see you. May I be of assistance?"

Marissa glanced around the shop, noticing it was somewhat crowded, and wondered why he paid special attention to her.

"Thank you, kind sir," she replied in keeping with the essence of the Faire, "but I am only admiring your handiwork. It is quite beautiful."

"Thank you, dear lady," he said, obviously pleased with the compliment. "Perhaps I can show the lady something that might be to her liking? Something that was made especially for you?"

Again, Marissa wondered at the special attention he was giving her compared to the other patrons of the shop. She also noticed that he kept bowing his head slightly when he spoke to her, as if he felt she was someone special.

"Oh, I suppose if it was made especially for me," she teased, knowing that couldn't possibly be true. "What could this fine smithy have fashioned especially for me?"

He grinned in obvious pleasure and said, "One moment, dear lady, I shall fetch it straightaway." And he disappeared through a rough-hewn wooden door to the back of the shop. He reappeared almost immediately, carrying a mannequin head made of metal, not Styrofoam, and draped with a chain-mail headdress that appeared to be made of gold. He set it reverently on the wooden counter, turned the head to face Marissa, stepping back and lifting his hand with a dramatic flourish, as if unveiling a great masterpiece.

She stepped closer to the work of art. It shone and glinted in the dim light of the wooden shed, little sparks of light dancing across the surface.

"It is beautiful," she breathed, but she knew it must cost a small fortune. Even if it wasn't real gold, and she was sure it wasn't, it was exquisite.

The shopkeeper's grin was genuine and he would have burst his buttons with pride, if his shirt had had any buttons.

"Milady," he said, drawing closer to Marissa and lowering his voice in a conspiratorial tone, "as I said, it was made for you. Please try it on, and you will see. It will fit you perfectly, I am sure."

"Oh, it is lovely," Marissa said, "but I could not possibly afford it. It looks like real gold, what is it made of?"

"It is gold, Milady. But you do not have to afford it, it is yours."

Marissa looked into the shopkeeper's eyes, not sure if she had heard him correctly. Surely he couldn't mean to give it to her for free.

"I don't understand," she said, confusion causing her fabricated speech pattern to lapse. "What do you mean, it's mine?"

"Exactly that, Milady. It was fashioned for you, it is yours. Please take it."

She looked from the shopkeeper to the golden headdress and back again.

"No, I couldn't possibly," she insisted.

He didn't try to hide his disappointment. "Perhaps I should continue holding it for you then, Milady?" he said.

She didn't know exactly what to say, so she acquiesced. "I suppose that will be fine," she agreed.

He nodded and picked up the mannequin head. "When you need it, come back for it. It will be here."

As he turned to go, a woman in a bright purple Renaissance gown, standing next to Marissa said, "I'd like to try that on, sir."

The shopkeeper looked at the woman and shook his head. "I am sorry, dear lady, but this is not for sale. I have something else that you may like," and he quickly disappeared behind the door to return the gold chain mail headdress to its hiding place.

"Well, I don't want something else," the woman said as the shopkeeper reappeared, carrying a lovely silver headdress. "I want that gold headdress. It's exactly what I've been looking for."

"I am sorry, madam," he said. "Perhaps I can show you something else."

"I don't want something else," she insisted. "I want the gold one."

"Again, I apologize," he said, "but as I said, it is not for sale."

"You were just about to sell it to that woman, and she didn't want it. Now I insist you sell it to me."

"No madam, you are mistaken. I was not selling it to the Lady Marissa, for it already belongs to her. I am simply holding it for her."

The woman in purple continued to argue, but Marissa had stopped listening. The shopkeeper knew her name. How could that be? She hadn't told him her name; she was sure she hadn't. Yet he'd called her by name. She looked down at Merlin who looked up at her with sublime, yellow eyes. She could swear he was grinning.

Marissa turned away from the shop and continued down the dirt thoroughfare, Merlin stepping majestically beside her. People milled about, talking and laughing. She marveled at the wonderful costumes and the way so many people got wrapped up in the magic of the day. She pondered the question of how the shopkeeper might know her name a few moments more, then dismissed it from her mind. It was unimportant, and she was enjoying the atmosphere too much to allow it to be tarnished by an unanswerable question.

After a moment or two, Marissa noticed several people watching her quite intently. Some passersby made complimentary comments about her dress, or the well-behaved cat at her heel, but there were intent stares coming from those dressed in the more authentic costumes.

A young woman dressed in a green, fairy costume with colorful veils hanging from her shoulders and waist chatted with two little boys that looked to be about five and seven in age. Holding tiny rocks of bright colors, she told the boys they were pieces of her homeland. She said that she came from a magic land and these were fairy stones that would make wishes come true.

When she stood up and turned, spying Marissa, she sucked in her breath in a surprised gasp. Smiling brightly, the young fairy woman curtseyed quickly, dipping her head in obvious reverence, and whispered, "Milady, you've returned. I knew the time had come. I will alert them all."

She then spun on her slippered heel and hurried through the crowd, with Marissa staring wide-eyed after her. 'The time had come?' What the hell did that mean? She stood dumbfounded for a few more moments, then realized that Merlin was tugging on his leash. She looked down and saw that he was trying to lead her in a specific direction.

"Ok, I'll bite," she muttered. "Where do you want to take me?" she asked, and allowed him to lead the way.

As they walked on through the throngs of people, Marissa wondered if Merlin was in on this joke. First the little shopkeeper knew her name, now this young fairy girl seemed to be expecting her. And she had the unshakable feeling that some of the people peering at her were not merely admiring her attire, but looking at her as if they knew her. An eerie feeling began in the pit of her stomach, and Merlin's behavior was only making it worse.

They continued down the dirt lane, stepping this way and that to avoid bumping into other Faire-goers, Merlin leading the way to a definite destination. They rounded a corner, Merlin tugging on the leash with absolute purpose.

"So, my little four-legged friend," Marissa said, "what do you have in store for me? Introducing me to some of the denizens of the Renaissance period who've time-warped to the present?"

He looked back up at her and meowed, never breaking his stride. Stepping purposefully up to a wooden counter, Merlin stood on his hind legs and placed his front paws against the wall, meowing up at the young woman behind the counter. She was dressed as a tavern wench with an extremely low-cut blouse, cinched very tightly for the purpose of thrusting her bosom impossibly high, and a full skirt made of a course muslin-type material. Marissa started laughing when she realized it was a food stand, and Merlin was trying to get a turkey leg from the young woman.

FOUR

"Okay, you little mooch," she said. "I guess we can have something to eat." Turning to the tavern wench, Marissa ordered a huge, roasted drumstick and a cola. As she took money out of the leather pouch strapped to her waist to pay for her order, the serving girl waved it away.

"No Milady," she said. "No need for money."

"What?" Marissa said, confused. "I'm trying to pay for my food."

"That is not necessary," the girl replied. "Here, this is for you," and she handed Marissa a turkey leg wrapped in a large square of waxed paper, and a plastic cup of cola.

Marissa took the food, and turned away, totally perplexed.

As she stepped away from the counter, the young girl said, "'Tis good to have you back again, Milady."

Looking at the girl, Marissa just nodded, then made her way to a large wooden table. She sat down, and Merlin jumped up on the bench next to her. Absentmindedly tearing off a piece of meat, she offered it to the cat, who took it eagerly, purring like a buzzsaw as he chewed happily.

This day was getting weirder and weirder. Marissa bit into the drumstick, not even tasting the rich, roasted flavor. She could not figure out why everyone seemed to know her. She could not figure out why she was so special to the characters here. She had been to the Renaissance Faire every year for many years now, and had never been singled out before. She had never been noticed beyond the everyday admiring glances she got for her costume, and the fact that she wasn't exactly ugly. But now, around every turn was someone who seemed to know her personally, someone who seemed to think she was someone special.

Marissa realized that every time someone called her 'Milady', they did so with a measure of awe and respect. It went far beyond the normal interactions of the Faire players and patrons; it seemed genuinely directed at Marissa specifically.

She continued gnawing on the turkey leg, absently feeding Merlin with pieces of the meat. Slowly she began to remember little things from past Faires. Every once in a while, as she admired the wares of the shops along the thoroughfare, someone had nodded to her in a familiar, almost conspiratorial manner.

As she sat at the wooden table now, she remembered several occasions when this happened. It seemed to her now that it only happened once or twice each year, but it did indeed happen each year. It always happened to be one of the Faire players, never someone dressed in regular street clothes. Each time, she remembered chalking it up to someone thinking she may be someone else, or just people being friendly.

Now, however, she realized a pattern, a definite flow to the occurrences. Oh, it wasn't necessarily in front of the same shop each year, or even by the same player, but the mannerisms were the same, the look of recognition was always there.

Marissa looked down at Merlin who now sat majestically cleaning his face with his paw. She watched him lick his paw over and over, then rub it along his cheek and nose, wiping away every trace of grease from the meat.

"Well," she said, petting the soft black fur on his head, "what do you think about all this? Is this some elaborate joke, or am I imagining things? I could have gotten what appeared to be a solid-gold headdress absolutely free. We got our lunch absolutely free. What do you think is going on here?"

He stared into her eyes, and again she could swear he was grinning at her.

"Okay, now you're freaky too," she said, giving his head one last rub. She stood up, threw the turkey bone in the large, metal garbage drum next to the table, and continued walking. After several steps, Marissa saw an old woman dressed as a crone. She wore what appeared to be a dress made of shredded, multi-colored rags, with a rag wrapped around gray, straw-like hair, and a large gap in the front of her brown-stained teeth. She fished a bright, red piece of thread from a colorful pouch strapped to her waist, giving it to the little girl she was talking to.

Marissa chuckled to herself about the dedication the players had to the Faire; purposely leaving your dentures home, and using God-knows-what to stain your teeth to play a role. Well, she supposed if you had it to give, why not?

The crone had handed the thread to the little girl and was telling her it was a magic thread when Marissa walked past her. As Marissa stepped past, the crone stood up straight and, the little girl forgotten, crouched into a shaky, unsteady curtsy.

"Milady," she murmured. She said nothing more, and the little girl looked at Marissa in curiosity. The girl's mother also looked at Marissa, then back to the crone, then back to Marissa. The mother seemed to sense something from the crone, and studied Marissa intently. Marissa met the woman's gaze and felt a shiver run up her spine. The woman, when looking into Marissa's eyes, seemed to shrink back, not from fear, but awe. Marissa realized that even this woman, dressed in plain, contemporary clothes, saw something more in her.

The crone bowed her head in reverence to Marissa, smiling a gap-toothed smile at her. She said nothing more, almost as if she felt unworthy to speak.

Marissa continued on her way, the eerie feeling in her stomach turning into a knot of tension. Now, not only were the Faire actors treating her as someone special, but the general public seemed to be acknowledging her in some way as well, and to make matters worse, with each step, Merlin seemed to become more and more cognizant of the special treatment.

Rounding a corner, Marissa almost walked directly into the leather-clad chest of the Black Knight. She stumbled, and he quickly grasped her elbow to steady her.

"Oh," she exclaimed. "I'm sorry, my mind is wandering."

"No matter, dear lady," he answered, his voice so deep it could be threatening, but he, too, spoke with respect. Immediately, Marissa realized she had never heard his voice before, and was surprised at her own reaction to it. She had always found the knight to be an attractive figure, mysterious and frightening, but now she could almost feel herself falling in love with him, as ridiculous as that was. She stepped back and shook her head to clear her mind.

"Please pardon my clumsiness," the knight said. "I would never forgive myself, had I harmed you."

Marissa smiled at him. It was her fault, and he was taking all the blame. Yes, he certainly did know how to win a lady's heart.

"Don't be silly, Sir Knight," she said in her best 'damsel in distress' voice. "'Twas I who was the clumsy one."

"Milady could never be clumsy," he stated, and bowed deeply at the waist. Marissa's heart skipped a beat, and she fought to control her breathing. Letting this wolf know how he was affecting her would be a disaster at best.

"Well, if you will excuse me," she said, taking a step backward to put a little space between them. She turned to resume her walk, when he again took her elbow.

"If Milady would wait just a moment," he said, his deep voice softened into a whisper. His grasp on her elbow was gentle, not at all threatening, yet she knew it was for a purpose.

"I would be greatly honored if you would allow me to accompany you to meet the others," he said.

She looked at his leather-clad face, into the dark blue eyes she could just barely see through the slits in the mask. Again, no threat, but a need, a purpose. She took another step back, and gently pulled her elbow from his strong hand. She had questions she wanted answered, but was not sure if this man would oblige.

"Tell me," she demanded.

He looked at her as if he didn't understand. "Milady?"

"I said tell me," she repeated. "You know what it is I want, now tell me."

He was silent for several moments, looking into her eyes. The fact that she could not see his face began to infuriate Marissa. He could read her every expression, yet she was limited to only his eyes. One can see many things in a person's eyes, but they are not the only means of expressing one's thoughts. She wanted to see his face.

She waited in angry silence while the knight stood rigid. Finally, he sighed and nodded.

"Yes, Milady," he said. "You do not appear to have all the information you need for your challenge. If you will follow me, I will offer you all I have."

She studied his eyes, wondering if this were a trick. If she went with him, would she be safe? She decided to take the chance and see what she could find out. After all, there were thousands of people at the Faire, what harm could come to her?

"Very well," she said. "I want to know exactly what's going on, and if you can shed some light on it, then I'll go with you."

Again, he bowed, and offered his arm to Marissa, holding a gloved hand out for her to place her hand atop. She laid her hand on his and allowed him to lead her between two wooden buildings. Merlin walked beside her, and she noticed he was not the least bit anxious. That gave her some comfort at least.

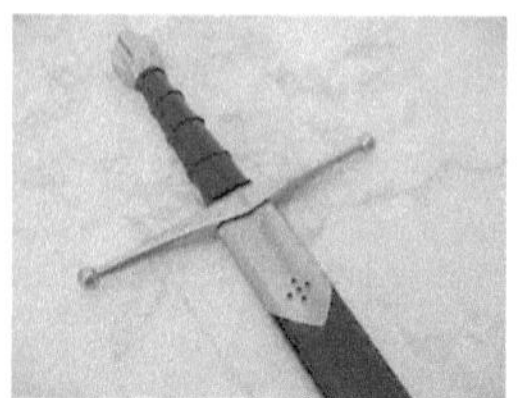

FIVE

The knight escorted Marissa into the back of a wooden shop. The room was sparse, with only a carved wooden chair against one wall. A door led to the front of the shop, and Marissa would have thought this room would be storage for the shopkeeper's wares, but there were none.

There were only three people in the shop beside Marissa and the knight; a serving wench dressed almost identically to the one at the food stand, with bosom thrust just as high; a short, round, shopkeeper wearing leather leggings and a white shirt with puffy sleeves; and a young woman dressed like an Elizabethan Lady in a jade-green velvet gown with a tight bodice, full skirt, long, cape sleeves that draped to the floor, a matching jade-green hat with a green feather tucked into the ribbon, and high-topped button shoes.

As Marissa and the knight entered, all three seemed to jump to attention. The wench curtsied and the shopkeeper bowed. The lady walked forward with confidence, yet great respect.

"Milady," she said, and grasped Marissa's hand. "We are glad you have come to claim your challenge."

Again, the word 'challenge', Marissa thought. What the hell is going on here?

"The lady wishes some information," the knight said. "I have offered to give her all I can."

"Yes," the Elizabethan Lady said, "of course. We will do whatever we can. Please sit down," and she led Marissa to the chair against the wall.

22

As she sat, Marissa said, "Please tell me what is going on here. Why does everyone seem to know me?" Merlin perched at her feet, looking every inch the majestic statue.

The Black Knight had backed up and now stood directly across the room from Marissa, his arms crossed in front of him. He seemed to want to relinquish control of the situation to the Elizabethan Lady.

"Everyone seems to know you," the lady said, "because they actually do know you."

"I don't understand," Marissa said. "How can everyone here know me? I've only been to the Faire one weekend each year. I've never met any of you before." She glanced at the knight. "I've seen some of you around, but have never actually met you."

He gave a short nod in agreement.

"No formal introductions have ever occurred," the lady agreed, "however, you are the one foretold. You are *she* about whom the Old One has prophesied."

The lady paused, giving Marissa a chance to absorb her words.

Marissa sat staring at the lady, trying to comprehend what she had just heard. Seconds passed, seeming like hours. Finally, she shook her head and stood up.

"I don't know what kind of joke you're trying to play here, but I'm done," and she marched toward the door. She tugged on Merlin's leash, but he dug his claws into the wooden floor and meowed loudly. Marissa stopped and looked at the cat. He stared up at her with pleading eyes.

"Oh, don't you get in on this act, you little feline traitor," she said, and gave his leash a tug. He meowed even louder, and tried to pull Marissa back to the chair. No one made a move to stop her from leaving, however, which was a relief to her. But she still wanted to flee this insanity, if only Merlin would cooperate.

"Please Milady," the knight said softly. "Heed the words Ruth has told you."

"The words Ruth has told me?" Marissa repeated. "She's told me madness. I'm the one prophesied? What kind of garbage is that in this day and age? You all seem to be taking your play-acting just a little too far. I've done some acting in community theatre myself, and it's great to become the character you're playing, but don't you think this is just a little ridiculous?" And she again tried to pull Merlin to the door.

"Milady," the shopkeeper said with urgency in his raspy voice, "please listen to all, then make your decision."

She looked at him, then looked at the Ruth. She then looked at the serving wench, who looked up at Marissa through her eyelashes with head respectfully bowed. Marissa shook her head again, confused, and looked at the knight for some explanation. He simply stared at her through the slits in his leather mask.

"Milady," Ruth said, "this is what was foretold. We have no more control over it than you. We are simply following our destinies. We need you; you are the one. If you refuse your challenge, we will be lost."

"Okay," Marissa said defiantly, "what is this challenge you keep talking about?"

Silence filled the room for several long moments. All eyes were on Ruth, as she seemed to be the one in charge of the little group.

Heaving a tremulous sigh, Ruth began.

"Milady, the prophecies have foretold of a foreigner coming to save us. Our queen grows old, and has no heir. There is already talk of fighting within her realm, and she is in great distress over this. We all wish for a peaceful ascension of our new ruler, but without an heir, this cannot be. However, the prophecies of the Old One can lead us in our quest for our rightful sovereign. No one would challenge the rightful heir to the throne, if she fulfills her destiny as foretold."

Marissa stared at the woman. Was she hearing correctly? This was all too much. This was something out of a fantasy novel. The Old One, the prophecies, Marissa's challenge. What in the world was going on?

Ruth continued. "The challenge, Milady, is that you must prove yourself worthy of the crown. Sir Erick will assist you in your training," and she nodded toward the knight.

Marissa's breath caught in her throat. Sir Erick? Why Erick? Why not any other name but Erick? George perhaps, or Ralph even, anything but Erick. Oh, these people certainly did have a sense of humor.

At the mention of his name, the Black Knight bowed slightly, his eyes almost black in the darkness of the room.

"No, no, no. This is all too much for me," Marissa said, and turned toward the door. She stopped short, though, in shock and wonder. When she had entered the little room from the back alley, there had been only one door, she was sure of it. Now, facing the door she had entered, she found that next to it stood another door. The two were identical, yet she was sure the second had not been there when she arrived. She turned back to the knight in confusion.

"Where did that door come from?" she demanded.

He looked at the set of doors, then looked back at Marissa.

"When we began to explain your destiny," Ruth offered, "the pathway opened for you. That is the pathway to your destiny."

Marissa stared at the woman. How had they managed that little trick, she wondered? Perhaps it was just a painted door. In the gloom of the little room, it would be fairly easy to paint a door to look real. She decided to get a better look, and walked toward it. All in the room seemed to hold their breath in anticipation. Even Merlin sat still as a statue, waiting.

Marissa dropped Merlin's leash and walked up to the door on the right, certain she had originally entered through the door on the left. She reached out to grasp the carved wooden handle, certain it would be a flat, painted surface. When her hand wrapped around the cool wood, she gasped in surprise. She drew her hand back, unsure if she should try to open the door. Finally, deciding to see once and for all what was going on here, she grabbed the handle and flung the door wide open.

Marissa stood in the doorway looking out. Since she had entered the little room from the alley between several shops, she expected to see the backs of other shops all running down the alley. Instead, what she saw awed and amazed her. Running out from the doorway lay a path, bordered on each side by vast fields of thigh-high grass. The path curved gently as it made its way into a wooded area.

Marissa stepped cautiously closer to the doorway to get a better look, and could not see the end of the tree line in either direction. Blades of grass waved and danced gently in an invisible breeze.

She stepped back and spun around to face the little group of actors. She had to keep believing them to be actors, for if she believed otherwise, her whole world would change forever.

"Explain this," she demanded in a voice she didn't recognize, and waited.

Again, silence reverberated through the darkness of the tiny room. Ruth then walked up to Marissa and grasped both her hands. Holding Marissa's hands gently, Ruth looked into Marissa's eyes and drew in a deep breath.

"Milady," Ruth began, "the prophecy foretold is coming to fruition. It is time for you to take your rightful place. It is time for you to face your challenge."

Marissa shook her head. "This is all a joke," she murmured softly. "This can't be real. Things like this just don't happen."

"Milady," the knight spoke up, "'Tis true. The kingdom is in need of you. You are the one foretold, only you can prevent the disaster from occurring, only you can save us."

"We will prepare you sufficiently for your challenge," Ruth continued. "You will have the necessary skills to prevail. Allow us to begin your training. You will see, it will all be natural to you."

With that, the shopkeeper shuffled to a wooden crate in the far corner of the room. As he opened the crate, Marissa could feel a palpable expectation in the air. Ruth and the serving girl held their breath, and the knight took one step closer to the crate, as if wanting to see something wonderful.

Marissa watched in silence as the little shopkeeper reverently lifted from the crate a sheathed sword with a golden hilt. The handle was carved and etched in a design Marissa could not make out from across the room, but was evident none-the-less. The sheath was fine, polished leather, with a dragon in battle tattooed along the length.

The shopkeeper turned, holding the sword horizontally with both hands in front of him. He cast his eyes down, looking at the sword as if it were a holy artifact. He turned toward Ruth, and offered the sword to her, but she shook her head.

"No, Terrence," she said gently. "Offer the lady her weapon."

Terrence then stepped timidly to Marissa, sword still held out in front of him, and bowed his head slightly.

"Milady," he said, his voice hoarse and soft with awe, "I present to you The Dragon Slayer," and he offered Marissa the sword.

She looked at the weapon. The etching on the handle resembled the etching on the sheath, of a dragon in battle. The sword looked immense, even though Marissa could tell it was not as large as the one strapped to the side of the Black Knight. The handle seemed just a little bit smaller, a little more slender, and the hilt curving outward looked to be made of solid gold.

"Wield it, Milady," Sir Erick said. "'Twas made for you."

Marissa looked at the knight, again hating that all she could see was his eyes. She swung her gaze back to the sword, thinking it would be much too heavy to lift, but reached out and grasped the handle anyway. The little shopkeeper released his grip on it, and stepped back. Absently thinking she would drop the weapon as soon as Terrence let it go, she was amazed that her grip held firm. The molded handle fit her grasp perfectly. The sword still lay horizontal, as that was how it was offered to her, so Marissa slowly raised the tip toward the ceiling of the little wooden shop.

It swung gracefully through the air, and stood at attention before her, almost as if boasting of its beauty. It amazed Marissa how light the sword felt. She expected it to be too heavy to lift, yet she hoisted it easily, as if it truly were made to fit her hand.

She grasped the sheath and drew out the blade, double-edged and looking sharp as a razorblade. It, too, seemed made of gold, as light glittered along both edges. But pure gold was extremely heavy, and this weapon felt light as a feather, she knew it must not be solid gold. She swung the blade gracefully before her, testing the weight and balance. It felt right in her hand, like it belonged there.

"Sir Erick is ready to teach you, Milady," Ruth said, breaking a palpable silence. "You will learn quickly, I am sure, as you were born for this."

Marissa stood in confusion, the sword balanced gracefully in her hand. Somewhere in the back of her mind she could hear a voice telling her to take what was offered. Somewhere in the back of her mind a voice insisted this was right, this was where her life was supposed to go. She had been searching for something to make her existence meaningful.

But common sense also voiced its opinion, and common sense told her this was ridiculous. Common sense told her this was all an elaborate joke. How could any of this be true? How could she possibly believe any of this was real? Her destiny was to become queen of some unknown land, to pass some challenge, and ascend a throne she had never known of until now?

This was all too much to bear. Marissa sheathed the sword and almost threw it back at the little shopkeeper.

"Hey, thanks," she said with some sarcasm, "but I'm outta here." She picked up Merlin's leash and turned toward the door to leave. As she faced the back wall, she stopped in her tracks. The second door had disappeared, and Marissa stood facing only one door. For a fleeting moment panic seized her as she wondered which door remained, but then she realized the placing of the door in the wall meant the one remaining was the one to the back alley bchind the shops at the Renaissance Faire.

She took a few steps toward the door, dragging the little black and white cat along the wooden floorboards as he meowed loudly.

"Please, Milady," the serving girl begged, "Avonridge needs you. You are our only hope to avoid disaster. Please, Milady."

Marissa stopped at the desperation in the young girl's voice. The pleading seemed genuine. She turned to face the girl, whose eyes were wide in anguish. Marissa looked from one face to the next, seeing the same anguish mirrored on each. The only face she could not read was the knight's, who still wore that damned leather mask. However, through the eye slits, she could see some great emotion, but whether it was fear, anger or humor, she could not tell.

"Look," she said to the little group, holding her hand up in front of her as if to ward off a hex, "I'm sorry about this, but your little play is over, at least as far as I'm concerned. Think of it this way, now you can go grab some other poor sap and try to convince her she's 'the one prophesied'. But for me, it's over. I'm going home, have some canned chicken soup for dinner, watch some T.V. and go to bed. I've had enough." And she opened the door and walked out into the alley with Merlin in tow.

SIX

It seemed like hours before Marissa finally fell asleep. She had lain awake replaying the events of the day, wondering why she had been handpicked to receive the brunt of the joke. It was well thought out by the players, beginning with the shopkeeper in the chain-mail shop, but it unnerved her greatly to think of the effort they had all employed to play this joke. It was so well contrived, so well-orchestrated. She had not recognized any of the players to be able to say some old friend, or foe, had staged it all. And she could not think of anything in her past that would prompt an old acquaintance to play such an elaborate trick on her.

Finally, after rehashing it over and over, she dropped into sweet oblivion, hoping her dreams would be soothing and comforting. Unfortunately, her hopes did not come to pass. Instead, she was visited all night by visions of Avonridge.

Marissa stood at the edge of a castle moat, the castle's drawbridge locked upright, preventing anything or anyone from entering or leaving. She stood with the golden sword in her hand, a golden shield with the crest of the dragon in battle strapped to her forearm, and wearing the golden headdress, along with a golden suit of armor. Beside her stood the Black Knight, also holding his sword at the ready. She felt tiny next to the large man, tiny and insignificant. But she knew in her heart she was not insignificant; she knew in her heart she was as capable as he. They were waiting for something, something terrible. Deep inside her consciousness, Marissa knew what was coming, and was prepared for it. Fear was not a part of the scene, but excitement bubbled and churned within the pit of her stomach. This was the culmination of her training; this was the event for which she had been prepared. This was her challenge. She must prove herself worthy, she must fulfill her destiny.

Suddenly the earth shook. She glanced at the Black Knight and saw him crouch slightly in anticipation. He too knew what was coming, he too knew the gravity of this challenge. Marissa must prevail at all costs, and the Black Knight would assist her in any way he could. He would stand beside her, fighting with her, but ultimately the duty was hers. It would be she who would defeat the something terrible that came this way.

The earth shook again, then again, as in time with footsteps. The knight swung his sword protectively in front of him as if to ward off the oncoming threat. She could hear his breathing, deep and measured, that of a trained soldier. She slowed her own to match his, forcing her hands to stay steady. The earth continued to shake with the rhythm of footfalls, extremely large footfalls.

Waiting for what seemed like an eternity, Marissa was patient. She knew the challenge was about to begin, and was ready. The footsteps grew ever closer, and from within the nearby forest, she could hear branches breaking. The something terrible drew near. Suddenly a roar launched from within the trees, a roar that abounded through the air. Marissa heard a woman scream from inside the castle walls, and she braced herself. Branches snapped within the forest, sounding like gunshots, and a tree came crashing down entirely. From the gap created by the felled tree, a huge, green-scaled foot emerged, tipped in talons as long as Marissa's sword.

Marissa bolted up in bed, sweat running down her face in rivulets. The sweat soaked her nightgown, as well as her pillow, and her heart pounded against her ribs as if to break free and fly away. Merlin sat perched at the foot of her bed, looking at her with calm yellow eyes. She scooted backward to sit up against her headboard, trying to catch her breath. The night was waning, and she could see pale light trying to force its way in past her flowered curtains.

She sat still for a long time, breathing deeply and forcing her heart to slow to its normal rate. The dream had been so vivid, so real. She felt as if she had truly been standing at the edge of a forest in preparation of a great battle. She had not seen anything more than the foot that had emerged from the forest, but she knew to what creature the foot belonged. So, this was her challenge. She was supposed to fight a dragon to prove herself worthy of becoming a queen. In the cool darkness of predawn, it seemed so clear, so logical. Perhaps she should consider the possibility that not all events of this world fell into the realm of what the general majority considered normal.

Merlin walked up to sit beside her, and she petted his head as he purred loudly. The routine of petting her cat calmed her greatly, and she laid her head back against the headboard, relaxing.

"So, this is it, huh?" she asked the cat. He had his head bowed, allowing her to scratch behind his ears, and didn't respond to her question.

"You think I should go along with all this nonsense, don't you?" she said. He continued purring, allowing her to answer herself.

"But this is ridiculous. I'm expected to fight a dragon?" she asked the little cat. "Exactly how am I supposed to accomplish that?"

She sat, petting Merlin's head in the dawning light.

"Suppose I don't succeed?" she asked. "Am I really considering the possibility of being killed? If I don't beat the dragon, I'll be dead. Is that something I want to just walk into?" She continued to pet her companion.

"This is all too much. The dream was so real, so logical. But, come on, fight a dragon? Really? Is my life so horrible that I'm willing to walk into something so final?" She closed her eyes and sighed.

"Courage."

Marissa opened her eyes and snapped her head up.

"What?" she demanded. She looked around the room for the source of the voice, but there was no one but the little cat. She looked at Merlin, who was looking at her with sublime, yellow eyes.

"Merlin," she said. "Did you hear that? That wasn't just in my mind, was it?"

In answer to her question, he simply continued to purr.

"Oh, I know," she said. "I'm just so on edge, I'm imagining things. I'm so amped up from the dream, I'm not thinking straight." Again, she laid her head back. "That dream has me thinking I can actually do it. Can you imagine?" she asked, staring at the ceiling. "Put my life on the line for a kingdom I've never heard of, for people I've never met before. My God, what am I thinking?"

"Strength."

Again, she snapped her head up, and this time looked directly at the cat. He was still looking her, sitting calmly.

"Merlin," she said. "What's going on here?"

Finally, the cat turned from her, and walked to the foot of the bed. He jumped to the floor, and then jumped right back up. He walked to Marissa, dragging something in his teeth, something shiny. When he approached his mistress, she tentatively reached out to take the shiny object from him. Her fingers grasped cool metal, and she jerked her hand back as if it had been burned. Sitting atop the quilt was the golden headdress from the chainmail shop at the Faire.

She stared at the headdress for several moments.

"How did this get here?" she finally asked. "You didn't carry it home, it's too big to hide. I would have seen it."

She sat quietly, contemplating all the occurrences of the past half hour. The dream, hearing someone speak, and now the golden headdress, all combined to destroy her resolve to stay away from her "challenge."

"Well, curiosity killed the cat they say, but what the hell? Perhaps we'll both survive. Besides, even if we don't, what do we have here? My life has become so mundane as to be almost painful. I have nothing here to make me stay. My job sucks, I have no one to share my life with, no offense, and the excitement factor in this world has all but evaporated for me. Why not see what this "challenge" has to offer?"

Merlin continued to purr, as if not understanding a word Marissa had said, but only enjoying the petting.

* * *

The Faire didn't open for another three hours. Marissa stood at the entrance in her Renaissance Lady gown, with Merlin on his leash. She knew the gates open at ten, but she also knew that if this was her destiny, she would be allowed in. And just as she thought, the Black Knight appeared beside her.

"Milady, I am glad you have come back. Please follow me," he said, and turned toward a door she had not seen before. She followed him through. The Faire grounds looked as they had the day before, only were now deserted. The knight stood inside the doorway, and when she stepped through, he held his arm out to her. She placed her hand atop his, allowing him to lead her down the dirt pathway toward the shop in which she had heard of her destiny. Merlin stepped easily beside her, acting as if this was the most natural thing in the world.

They reached the back of the little shop, and Marissa stopped.

"I have some questions," she said. "Would you please answer them for me?"

"'Twould be better if you let Ruth answer any questions you have, Milady," he said. "She can tell you everything you need to know."

Marissa grudgingly allowed him to put off her questions, and stepped through the door of the little wooden shop. She expected to see all three members of the group from yesterday, however only Ruth was present today.

"Milady," she said as she spied Marissa. "Thank you for returning." She took the golden headdress from Marissa, and said, "I know this must have been a difficult decision."

Marissa nodded, but said nothing, suddenly filled with a panic that she was making a huge mistake. She was on the verge of giving her life over to these people, people she had never met before yesterday. Was she really expecting to travel through a pathway to her destiny? Things like that just didn't happen in the modern world. This had to be some kind of joke, but Marissa couldn't figure out how the players could have pulled it off. She had seen the pathway into the woods. She had opened the second door in the shop and, instead of the backs of other shops, she had seen the pathway. Well, she had come this far, she might as well see it through.

"Come, Milady," Ruth said, "let us begin." She led Marissa to the corner of the room where the rough wooden chair stood. Marissa sat, and Merlin jumped up, perching protectively upon her lap.

"Now, I am sure you have questions. Please feel free to ask me anything."

Marissa stroked Merlin's silky fur, and nodded. "Yes, I need some things answered," she said. "First, am I to believe I am truly expected to become a queen?"

"Indeed," Ruth assured her. "Queen Gertrude has no heir, and has named you as her successor."

"But, why me?" Marissa asked. "I'm a total outsider, I don't even know Queen Gertrude. Why would she name me to be queen? Why not someone within her court? Why not someone younger? Do you realize I am pretty close to being past my childbearing years? The kingdom will have to go through this all over again when I name a new heir. Why doesn't she name someone else?"

"No one within her court is worthy, no one's destiny is to be Queen. That destiny belongs to you. It is not ours to question, we simply obey."

"But I know nothing of being a queen," Marissa protested. "I come from a society where there is no royalty. I've only read about it and seen it on T.V., I have no first-hand knowledge of how a queen lives."

"We will teach you all you need to know," Ruth replied easily. "You will not be thrown into a situation for which you are not properly prepared."

Marissa absorbed that for a few moments, then asked, "About my challenge, I assume I will be properly trained for that as well?"

"Of course, Milady," Ruth said. "Sir Erick will prepare you for your challenge. You will be fully trained in the arts necessary, as well as for the physical demands of ruling the kingdom."

"That can't be something you can accomplish in one day," Marissa said. "How long is my training supposed to be?"

"Oh, rest assured, you will be a quick study. You will understand as soon as your training begins, you were born for this, it will all come naturally. I have no doubt you can fully realize your potential in a day. As for the training for your royal station, that will come after your challenge, as your challenge is only the beginning."

Marissa's heart skipped a beat at those words. She would be facing the dragon after only one day's training with the sword. That would be tomorrow morning. And if she didn't prevail against the dragon, her little adventure would be over. She still didn't fully believe all she was being told, however, her dream had been so clear, so realistic. She found herself wanting this to be real, wanting to see what it would be like to be a queen. Of course, fighting the dragon was a definite drawback, but nothing worthwhile came easily. And, if her dream was truly prophetic, Sir Erick would be beside her. That was no guarantee of success, but it was a whole lot better than standing alone.

"Okay," she said, standing up and causing Merlin to drop to the floor with an annoyed yowl, "where do we begin?"

Ruth clapped her hands and laughed. "Ah yes, Milady, I knew you would not abandon us." Then she spun to face the Black Knight. "Sir Erick, would you prepare the training area?"

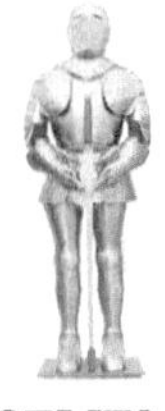

SEVEN

The Black Knight bowed slightly, and exited through the door into the back alleyway behind the shops. Ruth then walked to the front of the room and opened the door into the shop. The serving wench from the day before appeared, and behind her the shopkeeper from the chain-mail shop, carrying a golden suit of armor on a mannequin. Marissa gasped in awe at the magnificent sight, and realized that this was the exact armor of her dream. The workmanship was exquisite, the finish not smooth but hammered, each dimple of the hammerhead meticulously placed, and the gold shone like fire.

The shopkeeper bowed to Marissa, and stood the mannequin in the center of the room. He then turned around and disappeared through the door to the shop, and did not return. The serving wench went to Marissa, and removed her headpiece with the gold and black roses, which she handed to Ruth. Ruth hung the headpiece on a nail in the wooden wall, and the serving girl returned to Marissa to take her cloak. This too she hung on the wall, and the serving girl then began to untie the lace that ran down the bodice of Marissa's gown. She helped Marissa step out of the gown, and handed the gown to Ruth, who laid it carefully across the chair.

The serving girl then lifted the chain-mail headdress and placed it on Marissa's head, smoothing the cool metal down. She then took the suit of armor from the mannequin, and helped Marissa into it. Marissa marveled at how easily it slid onto her body. The golden breastplate fit snuggly to the waist, where it flared out with golden pleats to accommodate the cuisse that cover the thighs. Golden brassarts protected her upper arms, and golden gauntlets covered her hands and wrists.

Expecting the armor to be heavy and cumbersome, she was amazed at how light and uninhibiting it was. When she was fully suited, the serving girl called out.

"Terrence, Milady is ready."

Immediately the little shopkeeper emerged from the front of the shop into the little room. He opened the wooden crate and gently lifted the golden sword. Turning to Marissa, he held it in both hands.

"Milady," he said, "may I strap the sword on you?" She nodded mutely, and he shuffled closer to her. He grasped the leather belt and wrapped it around Marissa's waist, his head bowed in intense concentration. When he fastened the buckle and adjusted the sword to hang on her hip properly, he then returned to the crate and lifted from it a golden shield that looked absolutely huge, standing almost as tall as Marissa.

She gasped when she spied the shield, realizing it to be the exact one from her dream, with the crest of the dragon in battle. She had never seen the shield before, but in her dream she knew exactly what it looked like. The little shopkeeper then offered the shield to Marissa, and she instinctively slid her left arm through the leather strap on the back, and grasped the leather handle, holding it precisely the way it should be held. It too was deceptively light and comfortable, fitting snugly and properly. The two women then stepped back and admired Marissa.

"Milady," Ruth said, "it is wonderful to see you dressed and ready."

"But I'm not ready," Marissa replied. "Wearing the armor doesn't prepare me for battle."

"You are correct," Ruth agreed. "Sir Erick will prepare you for battle. You need not be apprehensive, this was all foretold, this is your destiny."

Marissa sighed. The panic she had felt when she entered the room was finally dissipating, and she began to enjoy the flow of the situation. The suit of armor fit like a glove, as if truly made for her. The chain-mail headdress was snug and comfortable, and somewhere deep within her mind, Marissa began to feel like this was right.

As if on cue, the Black Knight reappeared in the doorway.

"Milady," he said, his voice filled with awe when he saw her, and then seemed unable to continue.

"Yes," Ruth said, "Lady Marissa is ready for her training, Sir Erick. I am sure you will enjoy tutoring her, as she will learn quickly."

He bowed deeply at the waist, and said, "Milady, please come with me." He offered his arm to her again, and she placed her hand atop his, allowing him to escort her. Merlin, having somehow shed his collar and leash, trailed along behind.

The knight walked them between the backs of shops, still closed at this early hour, down the long alleyway. Shopkeepers were arriving and beginning preparations to open for the day. Here and there someone called out to Marissa, offering greetings of, "Welcome back, Milady," or "'Tis good you've returned, Milady." Marissa waved and smiled, almost feeling as if she actually knew these people.

Turning a corner, the knight led Marissa to the doorway of a large building. As they stepped through, Marissa recognized it as the auditorium in which she had seen one of the medieval plays the previous day. They had entered through the back and were up on the stage, but the building was empty save for the knight, Marissa and Merlin, who sat against the wall as if ready to watch a show.

Marissa's hand remained atop Sir Erick's, but he dropped his arm causing hers to fall to her side. She stopped, but he took two more steps, then spun to her with his sword raised in both his hands, slashing it downward as if trying to cut her completely in half. Her shield arm shot up and deflected the blow and she yanked her sword free, swinging it in an arc at the Black Knight's abdomen. He jumped back and cut downward with his weapon, driving hers to the floor harmlessly. She spun away from him and, with sword in the air, dove at him with a speed and ferocity that amazed and frightened her.

He deflected her sword and tried to knock her to her knees with his elbow. Again, she spun and with all her one-hundred, twenty-five pounds, shoved her shield at his side, knocking him off balance momentarily. Seeing her opportunity, she raised her sword and drove it at his side just beneath his arm. He brought his sword up to defend himself, but was a hairsbreadth too slow, and the tip of her sword connected with his leather armor.

Both Marissa and the knight stopped, frozen in stance. With Marissa's sword tip against his ribcage, his sword not able to deflect it, it was clear who was the victor.

Marissa took a step backward, and said, "How did I do that?"

The knight's deep blue eyes crinkled behind his mask, giving away his smile.

"This is your destiny, Milady," he said. "As Ruth told you, it comes naturally to you."

Marissa shook her head. "This can't be," she murmured. "I have never picked up a sword before, how could I just know this?"

The knight shook his head and replied, "It is not for us to know. We simply follow our destiny, wherever it may lead."

She looked at him, again hating that she could see only his eyes. So many things were falling into place, so many things were telling her this was right. The dream felt so real, she could feel the weight of the shield and the balance of the sword. And she had never seen the shield, but in her dream, she knew exactly what it looked like.

And when Sir Erick turned on her in attack just now, she didn't even have to think about what to do, her body just instinctively did what was necessary. And she had felt no fear, no momentary feeling of betrayal by the Black Knight, as if she already knew what would happen. It was his task to train her for her challenge, and she would learn quickly. Tomorrow she would face the dragon, and she would be ready.

Even though she knew she had to continue with her training, Marissa still needed to ask some questions.

"Tell me, Sir Erick," she said, "why must I face the dragon? Is there no one else who can fight it?"

He hesitated, and she persisted.

"If the dragon threatens the kingdom, why is everyone waiting for me? Cannot a knight such as yourself protect the kingdom?"

He sighed deeply and sheathed his sword. "Milady," he said, "I should not be the one to answer your questions, but for reasons of saving time, I will try. You are the Chosen One, the one to inherit the throne. Queen Gertrude has no heir, and so she named you to succeed her. It is said that a man in the kingdom wishes to usurp the throne, and since Queen Gertrude will die without children, it is possible for someone to lay claim to her crown.

Marcellus is that man, and he is wily indeed. Some say he is a wizard, while others say only that he is evil. It is said that it was Marcellus who first called Balvindor to attack the kingdom. If this is true, then Marcellus is truly a wizard, and no one can stand against Balvindor, no one except the one foretold. And that, dear lady, is you."

"But why?" she argued. "What gives me such power as to defeat a dragon simply because I was chosen?"

"Because it is so," he stated simply. He then raised his sword and attacked. Marissa easily deflected the assault, and on they practiced.

For several hours the two thrust and parried, lunged and sidestepped, attacked and deflected. Marissa grew increasingly more agile, more surefooted. Sir Erick had to teach her very little, she instinctively knew how to fight. Even the golden suit of armor seemed alive, it seemed to assist her in her movements instead of hinder her. On one particular attack from Sir Erick, Marissa was still recovering from a previous assault and did not see the knight's sword arc downward toward her shoulder. Even so, her arm was thrown up and the shield deflected the blow without Marissa realizing she was doing it. She stepped back and stopped.

Looking into the knight's eyes, she asked, "Who made this suit of armor?"

The Black Knight paused, and appeared slightly confused. "Why, the queen's goldsmith, Milady. Who else would make it?"

"The queen's goldsmith?" she repeated. "How could he know it would fit me?"

"It was not exactly made for you, Milady, but has been the queen's armor for generations. It has been given to you because you will be Queen."

She tried to think about that for a moment, but Sir Erick was taking his task of training her very seriously, and continued the swordplay.

EIGHT

After hours of fending off the huge knight's barrage of sword attacks, Marissa was beginning to tire. She could see Sir Erick likewise becoming fatigued, and felt grateful when he announced her training was complete. Suddenly the Black Knight ceased his assaults, brought his great sword up before his face with the tip pointing toward Heaven in a salute, and dropped to one knee.

"Milady," he said, his exhaustion only slightly evident in his voice, "you are ready."

An excitement gripped her heart as the words struck her ears. Even more thrilling though, was the fact that she felt absolutely no fear at all. She knew she would be fighting a dragon in the morning, but she didn't feel afraid.

She glanced over at Merlin, who sat still as a statue against the wall, watching the action with calm, knowing eyes. If Marissa let her imagination go, she could almost believe the little, black-and-white cat was a part of all this, that he knew what was about to occur and felt confident in his mistress. She sensed an odd feeling of support from the little creature, as if he believed she would live up to his expectations.

"Milady, it is time for you to rest now," Sir Erick said. "Ruth will see to your needs for the evening."

At that, Ruth appeared in the doorway, Marissa's black cloak draped over her arm. She walked to Marissa, and gracefully tossed the cloak about her shoulders, over the suit of armor. The gold armor shimmered within the darkness of the cloak, casting an awesome appearance. Even without a mirror to view her image, Marissa knew she was an impressive sight. Suddenly realizing her own thoughts, Marissa mentally kicked herself for her arrogance, and reminded herself that this still might all be a dream, or worse yet, a joke.

"Please follow me, Milady," Ruth said, and turned toward the door. As she followed, Marissa noticed the Black Knight bow to her, and she cast a grateful smile his way. Merlin dutifully fell into step behind her, and together they walked out into the early evening shadows. The Renaissance Faire had closed for the day, making the time later than seven in the evening. Marissa felt somewhat surprised to see how late it had become, she had practiced her swordplay all day, and felt only slightly fatigued. Under normal circumstances, since she was not exactly in athletic shape, she would have expected to be exhausted after a whole day of physical training.

She followed Ruth down the dirt path, and presently came to what looked like an inn set in Elizabethan times. Marissa had not noticed the inn at the Renaissance Faire during the previous day, and she felt sure she had never seen it at any of the earlier Faires. It was a large wooden building, with heavy doors barred from the inside. Ruth knocked three times, and a small window slid open, revealing the round, suspicious face of the innkeeper.

Ruth looked up at him and said, "Tis the lady," and the door was immediately thrown open. Marissa followed Ruth into the main room of the inn, Merlin at her heels. All eyes were turned to the lady in the golden armor, causing Marissa to bounce back and forth between anxiety and confidence. Each face in the room showed some measure of awe and respect. Even the huge Great Dane lying by the fireplace lifted his head and watched reverently as Marissa crossed the room. Merlin cast a defiant look at the dog, as if daring it to make a move, and strutted alongside his mistress.

The innkeeper led Ruth to a table next to the fireplace, and in dangerously close proximity of the Dane, making Marissa wonder if Merlin would be safe, but the cat didn't seem fazed in the least. He jumped up onto the chair next to Marissa's and began nonchalantly cleaning his face with his paw.

The Dane kept his eyes on the cat the whole time, and emitted one low, guttural growl. Merlin paused in his grooming and answered the growl with one of his own, the unmistakable deep cat growl that warns of immense pain if ignored, then continued cleaning his face. The innkeeper snapped his fingers once at the dog, and the dog lay his head down on his paws, and remained silent for the rest of the evening.

Marissa leaned her shield against the stones of the fireplace, and she and Ruth sat down. In moments, steaming bowls of stew were brought and placed on the table. Merlin stopped cleaning his face and looked expectantly at the serving wench, clearly wondering where his dinner was.

Spying the look on the cat's face, the serving girl gently patted his head and said, "Do not you worry little sir, yours will be along presently," and she disappeared through a door that led to the kitchen. Merlin seemed placated, and resumed his grooming.

In less than a minute, the serving girl returned with two drinking cups on a tray, and another small bowl, which she placed on the table in front of the cat. It was not hot stew, but a cooler mixture of raw meats, and Merlin obviously found it quite to his liking, considering the thunderous purr that erupted from his little body. The serving girl then place the drinking cups in front of Marissa and Ruth, curtsied quickly, and disappeared into the kitchen.

When the savory aroma of the stew hit Marissa's nose, she realized how famished she was, having eaten nothing since breakfast. She picked up the crude wooden spoon, and tasted the contents of the bowl. It was thick and rich, filled with a meat she could not quite identify, but found absolutely delicious. She wolfed down the bowl of stew, and looked around for the girl to see if she might get a second helping.

Ruth noticed that Marissa was finished with her stew, and motioned to the innkeeper. He opened the door into the kitchen and barked, "More stew!" Immediately the serving girl came rushing into the room carrying a steaming pot. She hurried to Marissa and refilled her bowl.

When she was finished, Marissa asked Ruth, "What kind of stew was that?"

"Mutton, Milady. Was it to your liking?"

"Yes, it was delicious," Marissa replied.

"Good. I had ordered the innkeeper to prepare it, knowing it is your favorite."

Marissa wondered how it could be her favorite, as this was the first time she had ever eaten mutton, but chalked it up to just another of the many mysteries of this adventure. She reached for the cup to take a drink. The cool liquid tasted like a heavy beer, and she wondered if she was also having her first taste of grog.

Merlin had finished his bowl of meat, and so resumed cleaning himself. Ruth had also finished her dinner, and the serving girl had cleared away the bowls, leaving the two women to sip their drinks.

Marissa began looking around at the other inhabitants of the inn. They were a group like something out of an old Robin Hood movie; huntsmen with daggers strapped to their waists; men dressed in dark leather like Sir Erick, but lacking his air of honor and chivalry; bawdy looking wenches with dresses cinched up so tightly their bosom could double as a serving tray. Marissa supposed some of the women were prostitutes, the men dressed in dark leather could be mercenaries, all in all, she wondered why Ruth would bring her to this place without the protection of the Black Knight.

"Time to retire, Milady," Ruth said, interrupting her thoughts. "Tomorrow holds much to keep you busy, you need your rest."

Marissa nodded and rose from her seat, strapping her shield to her arm. Glancing across the room, she realized that one of the possible mercenaries leaning against the wall had been studying her for quite some time. She had first noticed it when she sat down to eat, and all during her meal he had been watching her with a hot intensity. Now, when she rose, he turned slightly, as if preparing for something. Instinctively, Marissa knew what he was preparing for. She touched Ruth's arm to alert her, and Ruth turned to see what Marissa was referring to. She spied the man against the wall, then looked at Marissa with fear in her eyes. Marissa then knew she was correct about this man, and she also prepared herself for what would surely be occurring shortly.

Ruth fell behind Marissa, letting the lady in the golden armor lead the way, which Marissa gladly did. She made her way across the room, knowing that this inn would be providing her lodging for the evening, and headed for the staircase that was just a few feet away from the man. As she drew near, the man stepped away from the wall and unsheathed a sword Marissa had not seen before. In the blink of an eye, he was swinging the sword in an arc directly at Marissa's neck. Her shield arm shot up and deflected the blow, and she too unsheathed her sword. Just like in old western movies, everyone nearby immediately backed away, giving the combatants room, and saving their own skins.

After the day's punishment of continued swordplay, Marissa expected her muscles to protest loudly, but felt relieved to find them reacting as if she were still in training with Sir Erick. Her sword swung around in an attempt to cut her opponent in half at the waist, but he leapt back. He was not quite quick enough, however, and the tip of her sword sliced across his middle, putting a deep gash in his leather tunic.

Marissa wasn't sure if she'd caught flesh, but had no time to worry about it since he was attacking again. This time he aimed low, trying to cut her legs out from beneath her, but again she deflected, and now brought her sword down and caught his forearm, almost severing it. He dropped his sword and grasped his wounded arm, crying out in pain. Blood dripped onto the wooden floorboards as he hugged his arm to his chest.

Backing up against the wall, the pain in his face was overpowered by the hatred in his eyes. Marissa looked into those eyes and felt a deep fear. The hatred was intense and burning, and never having seen this man before, she could only conclude that his hatred was not for who she was, but rather what she was.

If she was to believe she was the Chosen One, as everyone had been telling her for two days now, then this man knew it and meant to stop her in her quest. And since she had never met this man before, it also seemed a logical conclusion that he would not be the only one. Now that she had accepted this course of action, was she going to have to battle every other man she would encounter? Was she going to have to fight human adversaries every day, whether or not she faced the dragon?

The man continued staring at her, and finally spoke.

"You will not succeed," he spat. "The Chosen One is no match for Him. You will see."

At that, the innkeeper lumbered up to the wounded man and, gripping him by the back of his collar, half dragged half carried him to the door. Opening the door, the innkeeper threw the man through.

"Your kind is not welcome here," he thundered. "Tell whoever sent you to stay away from my inn!" and he slammed the door, lowering the bar into the braces to lock it.

Looking around at the faces of the crowd, Marissa saw renewed awe and respect. Some nodded as if congratulating her on her victory, others looked as if they might applaud at any moment. Marissa didn't know if she felt sick, or exhilarated. She had never condoned violence, but this world was nothing like where she had grown up.

She had stepped into a world where one had to fight to survive, and this encounter was living proof. She had accepted a challenge that she knew would be dangerous; after all, she was agreeing to fight a dragon. She hadn't counted on having to fight men as well, but they shouldn't be as dangerous as a dragon, should they? A dragon could breathe fire, a dragon was as big as a five-story building, a dragon had talons and claws. A man was only a man, wasn't he?

Ruth stepped up to Marissa and took her elbow.

"Milady," she said softly, "I believe it is over for tonight."

Marissa looked at her. "Are you sure?" she asked.

Ruth nodded. "Yes, you have proven yourself. No one else will challenge you this night."

She followed Ruth up the stairs to a room at the end of the hall. A closet really, the room was about six by nine, and had a pallet on the floor covered in straw with two blankets thrown on top.

Ruth helped Marissa out of her armor, and stood it in the corner. Marissa took her sword and laid it next to the pallet, keeping it close at hand. The two women slid between the blankets, Merlin curled up between their heads, and in moments all three were asleep.

NINE

The sound of roosters crowing woke Marissa from a dreamless sleep. Somewhere in the back of her mind, she thought the sound was out of place, then she remembered. Already awake, Ruth sat in a corner polishing the golden suit of armor. Merlin stretched languidly, purring and rubbing his head against Marissa's shoulder. She reached up and absently scratched behind his ear.

Ruth noticed Marissa had awaken, and turned, smiling.

"Good morning, Milady," she said. "I trust you slept well?"

"Yes, thank you," Marissa answered.

Today was the day. Today her adventure would truly begin. She sat up and watched the intent young woman lovingly buff the golden armor. Marissa noticed a renewed reverence about Ruth as she rubbed the shiny metal with a soft cloth, as if she believed she was touching something mystical, something magical. She felt excited, but not afraid, and she wondered just how this 'challenge' would play out. If she were to believe her dream, it would be she and Sir Erick standing against the dragon, Balvindor. They would fight together, her teacher and she. She knew his moves, his tactics, and they would fight well together. They would be able to anticipate each other's actions, thereby giving Balvindor a much more difficult battle.

"You will have breakfast, then Sir Erick will come for you. Are you ready?" Ruth asked, but her voice said she already knew the answer.

"Yes, I actually think I'm looking forward to it. Funny, I'm not frightened."

"No need, Milady," Ruth replied. "You are prepared. When you are prepared for something, there is no need for fear."

"I suppose you're right," Marissa agreed. "I really feel like I can do this."

The fact that the dragon had a name gave Marissa a curious feeling. The dragon now had a personality, it made him a real entity. She would have to fight to the death, and killing something that had a name was different than hunting a wild animal. She knew animals had feelings, but they weren't thinking creatures, animals went simply on instinct. That this dragon had a name made it much more personal. She would be fighting a 'he', not an 'it'. That gave her an eerie feeling in the pit of her stomach.

Ruth smiled, then resumed polishing the armor. Marissa picked up the sword and studied it. The Dragon Slayer. The name certainly fit. The leather sheath had the depiction of the dragon in battle, as did the golden handle. The blade was about four inches wide at the hilt, double-edged and sharp as a razor. The balance was perfect; easily responding to Marissa's every command. This sword was made for one purpose and one purpose only; to kill the dragon. As she lifted the tip toward the ceiling, Marissa could almost feel the magic within. The sword seemed to speak to her, acknowledging her presence, reassuring her. With this weapon in her hands, she believed she could accomplish anything, including slaying the dragon.

Breakfast in the Inn's great room was not what Marissa had expected. With a battle to the death with a dragon awaiting her, she expected a hearty breakfast of meats and eggs. Instead, the serving girl placed before her a steaming bowl of something that looked like oatmeal. Its texture was similar, but it tasted more bread-like. Making it even more interesting, she found bits of meat mixed in. The whole thing made Marissa think of a sandwich, a hot, tasty sandwich.

As she ate the last spoonful, Sir Erick appeared in the doorway, as dark and imposing as Marissa remembered. Instantly she felt a surge of excitement, giving her a feeling of invincibility. Silently she reprimanded herself, knowing the easiest way to lose a battle was to become too egotistical. She knew her opponent would be formidable, and she needed all her strength and cunning. Being too sure of herself would only make for stupid mistakes, mistakes she could not afford to make.

The Black Knight strode across the room to stand before Marissa, and bowed deeply at the waist.

"Milady, the challenge awaits," he said ominously. "Shall we begin?"

She rose from the table, anticipation flushing her cheeks. She reached behind her and grasped the Dragon Slayer, strapping it around her waist. She lifted the shield and slid her arm through the strap, grasping the handle. Ruth walked up to her and placed the chain-mail headdress on Marissa's head. Merlin stood proudly at Marissa's feet, and meowed loudly. Sir Erick, Ruth and Marissa all looked down at the little, black-and-white ball of fluff and smiled.

"Yes," Marissa said, "it's finally time. Shall we go?" she asked the cat, and he meowed again.

"Milady," Ruth said, "before you go, I would like to wish you luck." The young woman moved and stood before Marissa. "We all are confident you will succeed. We know you understand the gravity of your undertaking, and you will prevail."

Marissa had not been nervous, until now. The sincerity in Ruth's eyes made her realize that, even though she was having a grand adventure, the fate of a kingdom rested upon her shoulders. Marissa had been going along with the players in this little production, not fully accepting that this might in fact be real. But the look in this young woman's eyes now made her realize that it most certainly was real. She could see confidence in Ruth's eyes, yet lying beneath the confidence she also saw fear.

She thought about the stakes here; Ruth's home was under siege by a dragon. She was relying on Marissa to save the kingdom; everyone was relying on Marissa. The weight of her 'challenge' just began to come to light. Marissa had been playacting, enjoying the attention. But now the play would be ending and the reality would begin. She would now be going to face a dragon.

Looking into Ruth's eyes, Marissa said, "I'll do my best for you, Ruth. You've been very kind to me, and I thank you. I won't let you down."

"I know you will not," Ruth replied. "This is your destiny, you can do no less."

"Will I see you again?" Marissa asked.

"Oh yes," Ruth answered. "I will be waiting for you at court. Queen Gertrude awaits the news. I shall advise her you have accepted your challenge, and await your return."

Marissa smiled, and hugged the young woman. She then turned toward the Black Knight, who bowed his head in silent acknowledgement, and led the way out the door with Marissa close behind, and Merlin trotting along at her heels.

Marissa walked beside the Black Knight, feeling her heart speed up with every step. Merlin trotted along behind her, completely at ease. Marissa still felt no fear, but was beginning to feel an incredible respect for the graveness of her circumstance. She would be doing battle with a dragon, presumably a dragon that could breathe fire and fly. She was the "Chosen One," the "One Prophesied," but she was still mortal. She couldn't fly, and she would certainly burn if she ended up in the direct line of a dragon-sized flamethrower. She had to keep her mind on her task. If she was to stand a chance against a dragon, she must use her instincts. After all, she found out while training with Sir Erick that her instincts were excellent for swordplay, as if she had been doing it all her life. She would now rely on her instincts to fight the dragon. If this was her destiny, it was not really in her hands. She must trust what fate was offering her.

Sir Erick led her to the path she had seen while in the back room of the little shop, the little shop where she had first heard of her destiny. When she looked out the door, the magical door that had appeared in the little room, the path had beckoned to her, called to her. She could feel the draw of what lay at the end of the path. The woods housed wonderful things, frighteningly wonderful things, and she would be encountering those things shortly.

As she walked, she could sense the suit of armor, feel its essence. It seemed to talk to her, reassure her. She could almost hear it. Even though it was made of metal, it seemed to breathe and move with her, as if a second skin. Not the least awkward or inhibiting, it actually gave Marissa energy as she moved. And as she and Sir Erick stood at the beginning of the path into the woods, the armor seemed to encourage her, seemed to push her onward.

The Black Knight turned to Marissa. "Milady," he said. "The challenge is yours. You must lead the way."

She looked at the path as it entered the woods. She looked at Sir Erick, who stared at her through the slits in his leather mask, his thoughts unreadable. She took a deep breath and stepped onto the path, beginning this journey that would forever change her life.

The thigh-high grass waved in the breeze. The Black Knight walked beside and a half step behind, deferring to the Chosen One. Marissa walked with confidence, Merlin scurrying along at her heels. The suit of armor continued to speak to her, to reassure her. It told her it would protect her, stand with her as the Black Knight would.

The cool darkness of the woods enveloped the unlikely trio; the lady in the golden armor, the Black Knight, and the little, black-and-white cat so full of innocent expectation. The path through the woods was wide enough for Marissa and Sir Erick to walk side by side, yet the knight continued to stay a half step behind. This was Marissa's challenge, Marissa's responsibility, and he was there merely to facilitate it.

She didn't know how, but each time the path branched off, Marissa knew which direction to take. She walked with a purpose, knowing where she was going, and eager to get there. Suddenly, she felt a leather-gloved hand on her shoulder. She acknowledged Sir Erick's warning, and slowed. About thirty feet ahead on the path stood a cloaked figure.

Dark and ominous, for a moment Marissa was reminded of the Grim Reaper, only this figure held not a scythe but a wooden staff.

"Marcellus," Marissa breathed, not sure how she knew.

"'Tis he," Sir Erick agreed softly.

At Marissa's heels, Merlin arched his back and hissed, all the hairs on his body standing out, almost doubling his size.

This was the man most feared by the group of people that had brought her to this pass. This was the man thought to be trying to take over the kingdom. An old man, he seemed ancient, yet Marissa could feel his strength.

"So, you have accepted your challenge," his gravelly voice snarled at her. "Why?"

His gray hair was long and scraggly, his beard draped down to the middle of his chest. He was exactly what she would expect to see wearing a pointed wizard's hat, but he wore a simple cloak with a hood. He was tall, almost as tall as the Black Knight, but was stooped and bent, and slender to the point of near-emaciation.

The genuine wonder in his voice amazed her, and she pondered his question a moment. She had asked herself the same thing several times over the past few days, and had drawn a conclusion.

"Because it is meant to be," she answered, her voice quietly defiant. "You will not succeed in your evil plot, Marcellus, you will not become king."

Marissa knew the hatred he had for her, she could see it in the way he stared at her, the venom emanating from his eyes, but she also saw his fear of her. He knew what everyone else knew, that she was the Chosen One, the one foretold, the heir to the throne, and he now could see she was here to claim her destiny. How disappointed he must be.

Suddenly Marissa realized the suit of armor was speaking to her, not with words but with thoughts, reassuring her. She could feel the energy of the gold, feel its strength. It seemed to be telling her she should not fear this man, that the armor was all she needed to stand against him.

"You are quite sure of yourself for being untried," he growled. "You think you will be able to battle Balvindor and win? I believe you are quite deluded. Why do you not abandon this fool's quest? You will only end up getting yourself killed, and the kingdom will be thrown into mourning for you as well as our dying queen. Save yourself the anguish of dying by fire, and the kingdom the pain of mourning for a fool."

"Your concern is touching, Marcellus," she said, her voice dripping with sarcasm. "I had no idea you were so thoughtful a man. But please, do not trouble yourself worrying about me. I am more than willing to accept whatever consequences my challenge brings. And as for the people of the kingdom, they need not mourn for me, as I will be just fine. Now, if you will excuse me, I have a date with destiny."

She resumed walking, the Black Knight a half step behind, Merlin at her heels, and growling softly. Her eyes never left the cloaked figure, and she watched as fury seemed to engulf him. As she drew near, her eyes locked on his, she noticed a change in his form, it seemed to waver, grow dim, then faded completely. Right before her eyes, the old man disappeared. She felt Sir Erick stop and heard his sharp intake of breath. Funny, she was not frightened by Marcellus. It was as if she already knew him, and knew his tricks. She continued on to her challenge.

TEN

The roar reverberated through the woods, making the ground shake. Marissa recognized the voice; she was finally about to meet Balvindor. It felt like she had been looking forward to this moment all her life. She felt slightly afraid, but the most overwhelming feeling was expectation. The fear actually seemed to calm her somewhat; if she felt no fear at all, she would have worried about her state of mind. At least she was rational enough to be afraid of fighting a dragon.

"How do you fare, Milady?" asked the Black Knight.

"I'm all right," she answered, ignoring the cold knot in her stomach.

Another roar shook the ground, and Sir Erick came up alongside her, walking shoulder to shoulder. Merlin growled at her heels and arched his back. Finally, the trio emerged from the woods into a large clearing. On the far side of the clearing stood the kingdom's fortress. It was probably a quarter mile away, but Marissa could see the drawbridge was closed, most likely to keep the dragon out. Knowing the initial battle would take place before the drawbridge, she led the way across the clearing. As they approached the fortress, the moat came into view. It was about seventy-five feet across, and seemed deep enough to swallow up the dragon.

Continuing on across the clearing, they heard another roar, this one sounding close enough to touch. Turning to face the oncoming attack, they stood as they had in her dream, side by side, Merlin off to the side, hopefully out of danger.

Her dream now played itself out in reality as Marissa stood at the edge of the castle moat, the drawbridge locked upright, preventing anything or anyone from entering or leaving. She stood with sword in hand, the golden shield with the crest of the dragon in battle strapped to her left forearm, and wearing her headdress, and the golden suit of armor. Beside her stood the Black Knight, dressed all in black leather, also holding his sword at the ready.

"So now it begins," she mused aloud.

"'Tis the hour of your challenge," he agreed.

She felt tiny next to the large man, tiny and insignificant. But she knew in her heart she was not insignificant, she knew in her heart she was as capable as he. They waited for Balvindor, and were prepared for him. Excitement bubbled and churned within the pit of her stomach. This was the culmination of her training. This was the event for which she had been prepared; this was her challenge. She must prove herself worthy, she must fulfill her destiny.

Suddenly the earth shook. She glanced at the Black Knight and saw him crouch slightly in anticipation. He too knew the gravity of this challenge, perhaps more than she as it was his homeland in jeopardy. Marissa must prevail at all costs, and the Black Knight would assist her in any way he could. He would stand beside her, fight with her, but ultimately the duty was hers. It would be she who would defeat the terrible dragon that came this way.

The earth shook again, then again, as in time with footsteps. The knight swung his sword protectively in front of him as if to ward off the oncoming threat. She could hear his breathing, deep and measured, that of a trained soldier. She slowed her own to match his, forcing her hands to stay steady. The earth continued to shake in the rhythm of footfalls, extremely large footfalls.

Waiting for what seemed like an eternity, Marissa was patient. She knew the challenge was about to begin, and she was ready. The footsteps grew ever closer, and from within the nearby forest, she could hear branches breaking. Balvindor drew near. Suddenly a roar erupted from within the trees, a roar that filled the air. Marissa heard a woman scream from inside the castle walls, and she braced herself.

Branches snapped within the forest, sounding like gunshots, and a tree came crashing down entirely. From the gap caused by the felled tree, a huge, green scaled foot emerged, tipped in talons that seemed as long as Marissa's sword. Now the dream was over and reality was about to begin.

Within the span of about half a second, the most fearsome creature Marissa had ever encountered materialized before her. A thousand feet away stood a thirty-foot-tall dragon. Funny, until this moment, Marissa had kind of doubted he was real. Common sense kept telling her that dragons didn't really exist, but she was staring at the proof that her common sense was mistaken.

Balvindor drew near with lumbering steps, and she studied him as he moved. He seemed so typical, almost as if purposely fitting her perception of a dragon. His head looked like a cross between a lizard and a bird; elongated snout with huge sharp teeth protruding from the mouth in both directions: up and down, with bony spikes sticking out around flat ears.

Set far back and high on the head, the eyes gave him a limited sideways view, but his neck was long and graceful, affording it great mobility to compensate for the lack of peripheral vision. His body was stout and muscular, and instead of short arms with useless hands, he had arms that reached out with agility and ended in almost human-like hands with long claw-like talons.

Protruding from his back just about where shoulder blades would be, were two graceful, transparent wings about twenty-five feet long and twenty feet wide. They seemed much too delicate to serve any purpose other than ornamentation. The short thick legs worked hard to haul the heavy body across the clearing, and the long feet with sharp talons stepped thunderously, making the ground shake with each step. A thick tail dragged along behind, leaving a rut in the clearing floor.

As Balvindor approached, Marissa watched him, assessed his strengths, which were many, and possible weaknesses, which seemed all too few. His thick body was covered in dinner-plate sized scales that seemed hard as iron, and she doubted that even the Dragon Slayer could penetrate them. As the dragon drew near, she could see that the only places not protected by the scales were around his arms, to allow for mobility, and between his fingers, which would not be very helpful unless the beast's heart was housed somewhere in his hands.

Marissa's mind raced, trying to plot her attack, thinking of a way to exploit any weakness she might find, but suddenly the thoughts of offense were swept away and defensive measures were called upon.

Balvindor was half way across the clearing when he suddenly leapt straight into the air, wings outstretched and flapping like a giant bird's. He lifted off the ground gracefully, transparent wings lifting him high over their heads. Sir Erick crouched and lifted his sword, trying to prepare for whatever the dragon had in mind. Marissa, however, saw the bend of Balvindor's neck as his head turned toward her, and his mouth opened wide. Instinctively, she threw her weight against the Black Knight, knocking him to the side, and brought her shield up, just as a burst of fire erupted from Balvindor's mouth.

The shield was almost as tall as she, and she crouched behind it as the flames came rushing at her. She saw the knight withdraw several feet to stay clear of the fire, and was relieved to see him safe, for the moment at least. Off to the side, over the roaring sound of the flames, Marissa could hear Merlin screech in fury, and desperately hoped he would stay out of the way.

The dragon's flame was heating up the metal of the shield, and Marissa wondered how long she'd be able to hold it without being burned. Suddenly the flame stopped, and the earth shook. Peering around the shield, she saw Balvindor had landed about fifty feet in front of her, and he was looking at her with intent eyes. She stood up straight, and immediately Sir Erick was at her side.

"Your reflexes are admirable, Lady Marissa," the dragon said, and Marissa almost dropped her shield in shock.

"You, you speak?" she stammered.

"Of course, I speak," he replied calmly. "How else do you think I communicate?"

"But, if you speak, that means you can understand as well," she said.

"I understand everything," he answered.

"And you know me?"

"I know you," he said. "You are the Chosen One. It is you who wishes to ascend the throne when good Queen Gertrude dies."

Marissa's head spun. A dragon that could talk, and knew her destiny, and spoke of the queen.

"You said 'good Queen Gertrude' as if you know her."

"I have lived in this land all my life," he said. "I know everyone of importance here."

"But you speak as if you respect Queen Gertrude," Marissa said.

"Great queen," he replied. "Generous heart, sharp mind, strong will. She will be sorely missed by all when her time comes."

"So, then, why do you terrorize her people?"

"Why?" he asked, as if he didn't fully grasp the meaning of the question. "I am a dragon," he said. "What else would I do?"

"I don't know. Why don't you try being friendly?"

"Friendly?" he repeated. "To whom would I be friendly? The townspeople who hate and fear me? I am a dragon, dear lady. It is not possible for me to be friendly."

When Marissa discovered the dragon had a name, it gave him a personality. But now, she knew he had intelligence and consciousness. That made the emotional battle all the more difficult.

Suddenly, he leapt across the distance separating them, reaching out with a huge clawed hand, trying to catch her. She drew The Dragon Slayer from its sheath and slashed at Balvindor's talon as it bore down at her head. Her sword made contact and bounced off, as if striking an iron bar. The hit was just enough to knock his hand away, and the talons closed on thin air.

Sir Erick rushed forward, driving his sword at the dragon's scaly arm, just as Marissa swung her sword up. The Dragon Slayer made contact with a dinner plate scale, and her previous assumption was proven correct, she would not be able to penetrate the dragon's armor. The force of the double assault, however, pushed Balvindor back a step, and he withdrew his arm.

Still studying her opponent, Marissa looked for any weakness she might be able to exploit, when her eyes came to rest on the soft, flexible area where his arms met his body. There were no scales here. These were the only vulnerable spots she could find besides the unprotected area just between his fingers.

Balvindor twisted to the left, and whipped his head around, striking at her like a snake, mouth open as if to swallow her up in one gulp, but Sir Erick leapt forward, driving his sword point at the dragon's eye. At the last possible moment, Balvindor spotted the sword and ducked his head, causing the weapon to glance off his bony skull. Immediately Balvindor's arm struck out again, trying to snare Marissa in his grasp, but she dropped to the side and his hand flew over her harmlessly.

Sir Erick slashed with his sword unceasingly, trying to drive the dragon back. Balvindor stepped back quickly, and whipped his tail at the knight, knocking him backward almost a full stone's throw. The Black Knight landed with a loud, dull thud, and lay flat on his back. Marissa feared for her devoted friend, but couldn't allow herself time to worry, as Balvindor was on the attack again.

With Sir Erick out of the way, the giant menace once again spun on the lady in the golden armor, snatching at her with a clawed hand. She brought the golden sword up over her head and stabbed at the palm of that hand as it whipped at her with amazing speed. The Dragon Slayer made contact with the palm, an area covered in smaller, less dense scales. The giant beast drew back momentarily, then turned to the side and whipped his massive tail about, trying to knock her to the ground the way he had the Black Knight, but Marissa was prepared and leapt over the tail like a jump rope.

Spinning around, Balvindor reached out again to catch her in his huge clawed hand. She raised her sword and drove it downward with all her strength. The razor-sharp blade came down between the last two fingers on the dragon's right hand. Marissa felt little resistance as the blade sliced cleanly through the flesh and bone. The last finger of the dragon's hand hung suspended in the air for a moment, then fell to the ground with a thud.

It seemed an eternity; Balvindor stopped, Marissa stopped, the air became still. The intelligent eyes of the dragon bore directly into Marissa's. He took a step backward, raised his head to the heavens, and shrieked out a howl of pain. Grasping his wounded hand in his good one, he leapt into the air. Graceful wings flapped and carried him out over the tops of the trees, and he disappeared into the forest.

Marissa stood and watched him go, then remembered Sir Erick. She ran quickly to the fallen knight, kneeling beside him.

"Sir Erick," she called, grasping his shoulder and gently shaking it. "Can you hear me?"

He moaned and rolled his head toward her. Finally, she saw him open his eyes through the slits in his mask. Immediately he sat up and reached for her arm.

"Are you well, Milady?" he asked frantically.

"Yes, I'm fine," she assured him. "Everything's all right now."

The knight hung his head. "Milady, I failed you," he said miserably. "You could have been killed because of my stupid error. I am unworthy to stand with you."

"Not at all, Sir Erick. Have you ever fought the dragon before?"

He looked at her. "No, never."

"Then it's something we both need to learn as we go along. You can't be expected to know what tricks he has up his sleeve until you've fought him a time or two."

She could see gratitude in his eyes, gratitude and shame.

"But, where is he?" Sir Erick asked. "Did you defeat him?"

"Well, not exactly," she replied. "But I did get in a good shot. I cut off his finger."

"Cut off his finger? Show me."

She stood up and hurried to retrieve her trophy. Picking up the amputated finger, she finally realized just how big it was. The smallest finger on the dragon's hand, and it was as long as her arm. She carried it back to Sir Erick, who looked at it in awe.

"Amazing," he breathed quietly, and they stood to proceed to the castle.

ELEVEN

"Open the gate for the Lady Marissa!" Sir Erick bellowed at the drawbridge. "Open the gate for the Chosen One."

Immediately the metal chains of the drawbridge began to clank and creak as the massive wooden gate slowly fell across the moat. At this close distance, Marissa could fully appreciate the strength of the drawbridge. Made up of lengths of lumber eight inches square, it looked sturdy enough to withstand any threat. And she certainly hoped it could, as it was the weakest point in the armor of the castle of the Kingdom of Avonridge.

The rest of the castle defenses consisted of the huge stone walls that surrounded the castle itself, and were topped by the parapets where the guards patrolled. The walls had to be at least fifty feet tall, tall enough to keep out most predators. Unfortunately, since Balvindor could fly, there wasn't much to do to keep him out.

As the drawbridge came to rest on the edge of the moat, the Black Knight gently took Marissa's elbow and gallantly escorted her onto the wooden gate. She walked with head held high, a purpose in her step. Having just bested a dragon, if only temporarily, gave her the confidence and strength she needed to continue in this adventure.

As they entered the courtyard, Marissa was awed. She didn't know what she had expected, but this was extremely impressive. The castle stood directly across from the drawbridge, approximately five hundred yards away. The length of the courtyard ran about twice the width, and was flanked by the huge stone walls of the fortress.

The castle itself was magnificent. A dozen stone steps led up to two huge oak doors that were flanked by immense stone pillars. The doors opened outward, and stood open now, making Marissa believe the castle was welcoming her.

To either side of the doorway, standing at attention were two men holding lances, with swords strapped to their waist. Dressed in leather tunics, it was obvious the two men were guards. Far above the doorway was the center-most of three stone spires, bigger than the outer two, approximately thirty feet across at the base and rising to a point about one hundred feet high. The top of the center spire pointed majestically toward the heavens, high above the walls of the fortress. Both side spires were lower than the center one by thirty feet or so, making the center the focal point of the castle.

The front wall of the castle spanned several hundred feet across, about triple the length of a football field. Windows were cut in the stone approximately ten feet off the ground and thirty feet apart, along the length. They were tall and very slim, five feet high by a foot across.

As Marissa and Sir Erick approached the steps to the castle, a young man emerged from the massive doorway. He appeared to be about fifteen years old, and was dressed as Marissa would expect a page to be; dark red shirt with billowy sleeves, dark brown trousers that were tight on the legs, but loose-fitting around the hips. His orange hair was long and combed straight back on his head, slicked down with something gel-like. Not at all handsome, his face made Marissa think of a mule, with long nose and jaw, and large teeth that protruded slightly from his very thick lips.

At the top of the stairs, the young man stopped and looked at Marissa, a look of joy evident on his homely face. He clasped his hands before him, and called out, "Milady, you have arrived! All is well now." He rushed down the steps toward Marissa, and dropped into a low, somewhat clumsy bow, almost toppling over in front of her.

The Black Knight reached out a hand to steady the youth, and said warmly, "Easy Jasper. The lady does not need you knocking her to the ground."

"Yes, of course, Sir Erick," the boy stammered, clearly flustered. "My apologies dear lady, please forgive me," and he bowed again, this time slightly more fluidly. "Please Milady, will you accompany me to be received by the queen? I have been sent to fetch you."

Marissa felt both relieved and anxious. She wanted to meet Queen Gertrude, but had never met actual royalty before and was nervous at the idea. Besides, not only would she be meeting the queen, she would be replacing her in the near future. That thought was disturbing. Since the main reason Marissa had been brought to Avonridge was to ascend the throne, that meant she would be anticipating the death of the current ruler. That thought just didn't sit well with her. Unfortunately, that was in fact the reason she was here, and she must accept it.

She walked behind Jasper as he nearly ran back up the stone steps to the castle doorway. Merlin strutted at her heel, head raised and looking quite regal himself, and Sir Erick followed behind. Jasper stopped at the doorway, looking at her with shining eyes, hopping from one foot to the other impatiently, but saying nothing for the sake of protocol. Marissa could see he was anxious and wished her to hurry, so she walked a little faster. As she reached the top step, Jasper turned and walked into the castle, leading the way.

When Marissa, Merlin and Sir Erick entered the doorway, they could see the young man several yards down the main corridor. He was looking over his shoulder and hurrying, and Sir Erick called out, "Jasper, show respect to the lady. Slow your pace and allow her to follow at a comfortable step."

Jasper immediately backtracked to them. Marissa felt sorry for him, clearly, he was trying to please Marissa, but also get her to the queen as quickly as possible.

"Forgive me, Milady," he said again. "I shall walk at a more suitable pace. Please follow me." And he turned and continued down the corridor at a normal stride. Marissa chuckled under her breath. She was beginning to get used to being treated like royalty, and was enjoying it completely.

The corridor was cool and well lighted. Torches hung on the walls and soft billows of smoke rose to the ceiling that seemed fifty feet high. Approximately twenty feet across, the corridor was wide, and several people were walking about. Everyone was dressed in elegant attire. Marissa marveled at the beautiful gowns with long, lacy sleeves that draped almost to the floor. The women all had head pieces that were made of delicate gauze in every color of the rainbow. The men wore tight shirts with big, billowy sleeves, and tight-legged trousers that ballooned out around their seat.

All seemed to know Marissa, and they each called a warm greeting as she passed. Not knowing anyone, Marissa simply nodded and smiled in response. She was sure she would be introduced to everyone in due time. Now her concern was meeting the good Queen Gertrude.

Jasper was hurrying again, unable to contain his excitement, and Marissa obliged by quickening her step. As they rounded a corner, Jasper stopped before a huge set of double-doors. To each side of the doors stood another guard, dressed much like those outside the main doors. This part of the corridor was deserted except for the guards. As the little group approached the doors, the guards, in unison, tapped their lances on the stone floor, making a loud click that echoed through the corridor. The group stopped, and, as one, the guards turned, stepped to the middle of the doorway, grasped the large, iron door handles and opened the doors in a fluid motion.

As Marissa watched the doors open before her, her heart pounded within her breast, making her catch her breath. The huge room was lighted by thousands of candles placed on oak tables. Cut into each side wall were four immense fireplaces, but only one on each side of the room was lit. Brightly colored cloth hung along the side walls, seemingly miles of it. Bright blue was met with bright red which was met with bright emerald green, draped gently along the entire length of each wall from the doorway to the far wall of the room where the throne stood.

And what a throne it was! Blonde oak polished to a shine, the throne was awesome. The back reached at least six feet high, and a carving of the dragon's head embossed it, a dragon's head that looked exactly like Balvindor's. The arms swept down and extended forward, curving gracefully to the floor. But the throne was nothing compared to the regal woman sitting beneath the carving of Balvindor's head.

Even if she were not sitting on the throne, Marissa would have recognized Queen Gertrude immediately. An older woman, appearing in her sixties or so, she possessed an aura of quiet power and strength. Since she was sitting, Marissa could only guess at her height, but she seemed to be about as tall as Marissa herself. She was of medium build, not at all frail looking, but having what Marissa liked to call 'substance', neither too heavy, nor too thin. Thick and dark, her hair held some gray streaks throughout, giving it a sophisticated appearance. It was combed straight back and hung in a thick cascade behind her shoulders. Atop her head sat a simple gold crown, with a dragon's head carved on the front.

Queen Gertrude's gown was also simplistic and elegant. Pure white to the point of luminescence, the bodice fit tightly to a small waist, the skirt billowed out around the legs of the throne, and the sleeves draped gracefully to the floor. The material looked to be satin or a similar type, and added to the regal air of the Ruler of Avonridge.

As Jasper disappeared, Marissa took a deep breath and began walking forward. The queen's eyes fell upon her, and Gertrude smiled, her face lighting up with genuine joy. Marissa's legs felt like lead weights as she made her way to the throne, not at all sure how this meeting would play out, but feeling reassured by the presence of Sir Erick and little Merlin, who trotted at her heels with all the authority and arrogance of a being that belonged in this great castle.

When Marissa was ten feet or so away from the throne, the queen stood up, a slight tremor just barely visible in her stance. With eyes shining, she reached out her hand, and Marissa obediently stepped up to grasp it.

"You are finally here," Gertrude said, her strong voice surprising Marissa. She had heard many times that the queen was ailing, but saw little evidence of that.

Marissa dropped to one knee before the queen and said, "Majesty, it's good to be here," and somewhere deep inside she felt exactly that.

The queen let go of Marissa's hand and returned to the throne. "Rise, my child," she said, and Marissa obeyed. "My information was correct, I see. I was filled with such joy when they told me you had finally arrived."

"Yes, Majesty," Marissa replied. "I understand you have been waiting for me."

"For many years," Gertrude agreed. "When you were born, I knew our prayers had finally been answered. And now all will be well."

Marissa didn't quite understand, but decided not to question things just now, her head was spinning already. Merlin purred loudly at her feet, and Queen Gertrude looked down at the little black and white ball of fluff with affection.

"Little Merlin," she said, "it is good to see you as well. It has been quite a long time, has it not?"

Merlin meowed loudly in response, and leapt up onto the arm of the throne. Marissa could hear him purr as Gertrude petted him and scratched his ears. She hoped he wouldn't jump into Gertrude's lap and cover her beautiful white gown with black cat hairs, but the queen laughed delightedly.

"Yes, you have been missed dearly," she cooed to the cat. "Now that you have returned, we can get on with the business of the kingdom." Again, Marissa was completely confused. The queen was talking as if Merlin was human, and she knew him.

Then Gertrude turned her attention to the Black Knight. "Sir Erick, I trust everything went well with the lady's training?"

Sir Erick bowed deeply at the waist, and said, "Yes, Majesty. She learns quickly, and handles the Dragon Slayer as if she were born for it."

"And indeed, she was," Gertrude remarked. "Splendid, splendid. Then everything is progressing as foretold. This puts an old woman's heart at ease."

Off to the side Marissa noticed movement, and Ruth stepped out from behind a curtained doorway. She was glad to see her friend, and turned to welcome her. Ruth walked to Marissa and dropped into a graceful curtsey.

"Milady, it is good to see you," she said. "Obviously you have met your first challenge, and prevailed."

"Yes," Gertrude said, "please tell us about your first encounter with Balvindor."

"Balvindor?" Marissa repeated, almost having forgotten the battle with the dragon. "Oh yes. Well, it was very exciting. I don't know exactly what I had expected, but I wasn't really prepared to meet him."

"Meet him?" Gertrude said. "What do you mean?"

"Well, I certainly didn't expect a dragon to be able to speak. But he talked to me."

"Of course," Gertrude said. "How else did you expect him to communicate?"

"I hadn't really thought about it," Marissa admitted. "I suppose I didn't really expect dragons to communicate. I mean, I guess I simply expected him to terrorize people and eat them, not talk to them."

"Well, as dragons go, Balvindor is quite civilized," the queen remarked. "I suppose we should feel lucky to have such a sophisticated dragon in our kingdom. After all, I have heard all sorts of stories about other kingdoms being terrorized by other dragons with very little class. Really quite sad in some instances." She leaned her head in her hand and thoughtfully tapped her chin with a beautifully manicured finger.

Marissa just stared at Gertrude, not knowing what to say. The queen seemed to speak of Balvindor with affection. A dragon being civilized? Sophisticated even? She would have to revamp her whole way of thinking in this beautiful, magical, confusing place.

"Ruth," Queen Gertrude said, turning to the young woman, "will you escort the Lady Marissa to her chambers and prepare her for the celebration?"

"Of course, Majesty," she replied. "This way, Milady," and she turned toward the curtained doorway through which she had emerged.

Marissa bowed to Queen Gertrude, in her armor she didn't feel a curtsey was appropriate, and hurried after Ruth, Merlin trotting behind.

TWELVE

Ruth led Marissa down a long, stone corridor, quite similar to the one she had marched down to meet the queen. Torches lighted the way, and cool breezes kept the air comfortable. Marissa wondered if the immense castle was drafty in the winter. With the ceilings so high and rooms so big, she wondered if the fireplaces would be enough to ward off the cold.

Ruth stopped at an oak door and announced, "Your chambers, Milady." She opened the door, and Marissa stepped into splendor she had never before experienced. The huge room had two fireplaces cut into the stone walls, one on either side of the canopied bed. No fire blazed, as the room was moderately warm. The bed was also quite large, with pale-yellow gauze draped around it. An ornate coverlet lay spread over the mattress with delicate swirls in pale-yellow, soft-pink and sky-blue.

Windows were cut in the stone high off the floor, the same windows as on the outside of the castle; tall, very narrow openings that allowed soft breezes from the warm autumn day to waft through. On one side of the room, next to a fireplace, stood an oak chest of drawers, and on the other side of the room, next to the second fireplace, an oak dressing table with a large mirror.

"I will have a bath prepared for you, Milady," Ruth said. "But first, would you like to rid yourself of your armor?"

Surprised at how comfortable the armor felt, Marissa had almost forgotten about it. Gently, reverently, Ruth lifted the chain mail headdress and walked to the chest of drawers, where a carved wooden head sat. She placed the headdress over the wooden head, then returned to help

Marissa remove her armor. Ruth gently placed the armor in a corner, then pushed aside a soft-pink curtain to reveal a closet recessed into the stone wall. She reached into the closet and removed a pale-blue silk dressing gown that hung on a crude, wooden hanger. She helped Marissa into the dressing gown, then turned toward the door.

"I will summon the bath now," she said. "You may wish to rest for a few moments," and she disappeared out the door. Marissa looked around the magnificent room, her eyes falling upon Merlin who lay on the bed, paws stretched out in front of him, head held high, watching her intently.

"So, my little friend," she said softly, "you seem to know exactly what's going on here. Mind letting me in on it?"

His gaze never wavered, but he opened his mouth and meowed. Marissa's heart skipped a beat, when, deep within the meow, she could swear she heard a human voice say, "Patience."

She stepped closer to him, looking into his yellow eyes, sure they were changing slightly. The irises that had always been completely feline, seemed to be rounding up a little. Instead of being the narrow up-and-down of the cat's eye, they were now taking on a distinctly human character.

"Merlin," she said, "are you something more than you've seemed all these years?"

He said nothing, but cocked his head slightly to one side, as if studying her as she studied him.

Suddenly Marissa doubted what she had been so sure of just a moment before. Merlin's eyes were completely feline again. She shook her head in confusion.

"I know what it is," she said to herself, as she turned away from the little, black-and-white cat. "It's simply my exhausted mind playing tricks."

"Are you sure?"

She spun back to face Merlin, who had not moved a muscle, but lay regally on the bed, head still cocked to the side. However, his irises were now fully human.

"Okay, now," she said, "if something's going on here, please tell me." She was amazed at the fact that this did not frighten her, but merely made her curious.

But Merlin continued to gaze at her in silence. She stared back at him, curiosity turning to deeper questions. Marissa could remember when she'd found the little cat. He was small, but obviously fully-grown. He had been sitting on the hood of her car one afternoon when she left work and, as she approached the car, he stood up and meowed, then began purring like a lawnmower.

She scratched him behind the ear, and he rubbed his head against her hand lovingly. When she opened her car door, he jumped in as if he belonged there. Marissa realized with a start that that had been fifteen years ago. During that time, Merlin had shown absolutely no signs of aging. He had no gray hairs around his nose, no slowing down as if from joint pain. It was as if he hadn't aged in the least.

"Who are you, Merlin?" she asked.

His only answer was to begin licking his paw and washing his face. She chuckled at him, not annoyed, but amused.

"So, you aren't going to tell me? Okay. I have a lot of thinking to do about this whole situation anyway, I guess you're just one more facet of it. Queen Gertrude seemed to know you personally."

At the mention of the queen, Merlin looked up at Marissa, his eyes still completely human in appearance.

"You do know her, don't you?" Marissa said. "And she knows you."

She was about to continue with her questioning, when Ruth entered the room followed by two large men pulling a carved, wooden bathtub behind them on a rolling pallet.

After dismissing the men, Ruth helped Marissa into the bathtub, then retrieved a pile of clothing from the curtained closet and laid it on the bed. When Marissa was sufficiently clean, Ruth helped her out of the tub and buffed her dry with a huge, cotton towel. She then helped Marissa into the exquisite satin gown that fit as if it had been made especially for her.

The gown was a deep-emerald green, tight-fitting at the waist with a low-cut bodice to show off a respectable bosom, large puffy sleeves that tapered down to snug forearms, and a billowy full skirt that reminded Marissa of the hoop skirts of the Old South. Ruth then had Marissa sit down at the dressing table where she fixed her hair and jewelry. As Ruth placed a necklace around Marissa's neck, Marissa gasped. Hanging at the end of a delicate gold chain was the largest emerald she had ever seen. Being approximately the size of a half dollar in diameter, it glowed a fiery green, and seemed to have a life of its own.

"This is gorgeous," Marissa breathed.

"Yes, the queen does have magnificent taste in jewels," Ruth answered.

"Does this belong to the queen?" Marissa asked.

"Of course," Ruth said lightly. "Everything in the castle belongs to the queen."

"Of course," Marissa echoed faintly. "So, she picked this outfit for me then?"

"This is the attire she wore at her arrival banquet. I suppose it is the tradition."

"Were you at her arrival banquet?"

"No, Milady. I was not born yet," Ruth replied.

"Her arrival banquet," Marissa repeated. "Did Queen Gertrude come from another land as well?"

"Yes, Milady. All the queens of Avonridge come from another land. It is the way of things here."

"That seems strange," Marissa said. "Doesn't the queen's child usually ascend the throne?"

"Our queens do not have children. There is such danger in childbirth, we do not wish to put our queens at risk."

"But don't they marry? Don't their husbands wish for children?

"No, Milady, our queens do not marry." Ruth's statement was so matter-of-fact as to put a close to the conversation.

Marissa thought about that for a moment. She had concluded that she probably would never marry, and so was not particularly upset by that thought.

Attaching emerald earrings that matched the fire of the necklace to perfection, Ruth stood back and said, "Ah, Milady. You look simply beautiful."

Marissa looked into the mirror, and her reflection pleased her. Her soft brown hair was piled on top of her head in a delicate bun, wrapped by an emerald studded comb that gave the illusion of a tiara. Ruth had outlined her eyes in kohl, which made them seem larger and more luminescent. Her cheeks held a hint of color, making Marissa seem several years younger than her actual age of thirty-seven.

"Shall we go?" Ruth asked, as she stepped back to allow Marissa to rise.

"Yes, I'm looking forward to this."

Merlin, who had been sleeping the entire time, finally stood up and stretched, arching his back like the typical black cat of witch's lore. He jumped to the floor and lazily walked to Marissa, looking up at her with human eyes. He meowed loudly, and Marissa could swear she heard him say, "Beautiful." She smiled down at the little, black-and-white ball of fluff, and the three walked out of the room.

THIRTEEN

At the sight of the lady, the immense room grew silent. All eyes were on Marissa as she walked straight to the throne and dropped into a curtsey, low and graceful. Queen Gertrude's face lit up in a lovely smile as she extended her hand to Marissa. Marissa rose obediently and grasped the queen's hand, and Merlin leapt up to the arm of the throne and meowed loudly. Gertrude laughed and patted the little cat.

The queen looked absolutely stunning, wearing another white gown, but this one much more elaborate, with a modestly low neckline, and glittering diamonds encrusted around it. The tight bodice was the same shimmery satin, but with diamonds enhancing its glitter. Atop her head sat a crown made of gold with purple velvet in the shape of a dome. Embroidered in gold stitching on the velvet was sewn the shape of the dragon's head that adorned the throne. From the regal elegance of the queen and the guests in the banquet hall, Marissa realized this "arrival banquet" was certainly a big deal.

Conversation returned to the room as those at the celebration approached Marissa to offer their greetings: "Welcome, Milady," "We're so glad to have you with us, Milady." Men and women dressed in fine party clothes, all obviously high-born of the kingdom, and all obviously pleased with Marissa's arrival.

Looking out over the crowd of revelers, Marissa spied a familiar figure. Sir Erick stood against the far wall, dressed in his usual black leather, arms folded across his chest, looking quite formidable and aloof. He seemed only slightly ill at ease, but when their eyes met, he visibly softened. She could feel his support from across the expanse of partygoers, and was grateful for it.

Sensing some expectancy, perhaps even urgency, to this celebration, Marissa began to feel somewhat nervous. She was looking forward to spending time with the queen, learning everything she needed to know about the kingdom and its affairs, but the enormity of the task was beginning to weigh on her. Gertrude seemed in fine health, so Marissa was sure she'd have plenty of time to familiarize herself with palace protocol, but the idea that she was slated to become the next queen was mind-boggling.

She approached the throne and said, "Your Majesty, may we talk?"

The queen smiled at her, but said, "Now is not the time for conversation, my dear. Now is the time for chitchat and fun. Go, walk among the people and enjoy yourself. This is your evening."

Wishing to sit and talk with Gertrude, but not willing to argue with her, Marissa did as she was told and turned toward the crowd. Immediately a man wearing a green velvet waistcoat and matching stockings, walked up to her, his hand extended.

"Lady Marissa, please allow me the honor of the first dance with you," he said, his voice warm and inviting. His hair was dusty blond, slicked back from his forehead and extending just beyond the collar of his waistcoat. He had an extremely nondescript face, nothing uncommon or unusual about his countenance. He was about six feet tall and slender, all-in-all, pretty much the garden variety gentleman.

She smiled and placed her hand in his, allowing him to lead her to the middle of the room. The crowd parted to let them pass, and Marissa's eyes fell on Sir Erick once more. However, this time his manner was not warm, but tense. He straightened up, and she believed she saw him scowl under his leather mask.

Stopping in the middle of the room, the man leading her faced her and placed his hand on her waist. Immediately, music erupted from the corner of the room where a group of six men sat playing stringed instruments.

"You have me at a disadvantage, sir," Marissa said as they twirled around the room. "You know my name, but I don't know yours."

He smiled, and Marissa grew slightly uneasy. "Forgive me, dear lady, my manners are atrocious. My name is Forsythe. You will have no trouble remembering me, as I am the man you are going to marry."

Marissa's unease blossomed into a ball of ice in the pit of her stomach. Forsythe's smile didn't waver, but instead of the warm welcome she had felt earlier, she now felt something completely different; his smile was calculating. Glancing around the room, she caught sight of Sir Erick, whose eyes did not leave her. He still stood against the wall, but instead of looking like he was in a festive mood, he looked ready to spring into action.

Marissa then spied Ruth standing near the queen. Both women were watching the couple dance, and Marissa could sense anxiety from them as well. She turned back to Forsythe, who was still smiling at her, or was it leering, and tossed her head back, laughing.

"As flattering as that thought is," she said lightly, "the Queen of Avonridge doesn't marry. So, I'm afraid that won't be happening."

"Simply because previous queens have not chosen a husband, does not mean you cannot," he replied. "You obviously have more character and strength than queens past. You can see the vast advantages to having a man to stand beside you. A man can take charge of the trivial aspects of ruling the kingdom, while you concentrate on affairs of state. After all, there is a reason it is called a *king*dom." Again, that leering smile.

A shiver ran up Marissa's spine. Having worked in a "man's world" her whole life, she had become attuned to the attitudes of the chauvinistic. Believing a woman could not possibly handle the duties and responsibilities of anything important, men felt they needed to guide or control them. Well, even if Marissa were of a mind to marry, Forsythe would certainly not be on the list of possibilities. Far from it, indeed.

"So, Forsythe," she said, masking her contempt behind a sweet, lilting voice, "should I consider your offer, what would the queen's husband be called? After all, there doesn't seem to be a previous example. Would you be a prince?"

He chuckled in a self-deprecating way and shook his head. "Of course not, my dear. As the queen's husband, I would be King." His eyes, a yellowish color, seemed to bore into her, almost as if he felt he could persuade her by will alone.

"I see," she murmured. "And, as King, you would be content with taking charge of the 'trivial aspects of ruling the kingdom', while the queen handled the affairs of state?"

His eyes narrowed, obviously sensing the trap Marissa was setting. "I would still be your servant, Milady," he said guardedly. "After all, you would be Queen."

"Yes, I *will* be Queen," she emphasized. "And apparently I have to deal with potential usurpers already." Her voice turned hard and icy. "Do not assume I can be easily influenced, Forsythe. I may be naïve, but I am not stupid."

He stepped back from her, his face an angry red. "I assure you, dear Lady," the word dripped with venom, "I have no intention of committing treason. Your accusation is highly insulting, and, as you are not yet Queen, you would do well to keep in mind that no one knows what tomorrow holds. Things have a way of coming undone at the last minute."

He turned to go, but her voice stopped him. "If I insulted you, that is unfortunate," she stated, "but you would do well to mind your tongue. Unless you know something no one else does, you're in danger of making an enemy. It seems to me that is a very foolish mistake. And if you do know something no one else does, my favor would be of little consequence to you. So, tread lightly, Forsythe. I have a very good memory."

His eyes blazed, but his face paled slightly. He bowed quickly, a small and insincere gesture, then walked away. Marissa caught sight of Sir Erick, and could swear he was smiling under his leather mask.

As she turned away from Forsythe, she turned directly into Marcellus. He stood grizzled and stooped, his gray beard hanging lifeless down his chest. His cloak was draped around his shoulders with the hood over his head, and his eyes gleamed out of the darkness and bore into her with a vehemence that was almost palpable. He loomed over her, his gnarled old hand wrapped tightly around his wooden staff. An aura seemed to emanate from him, an aura of dark power. It swirled around him, bathing him in cold anger.

She had felt a chill deep in her stomach, but thought it was only her reaction to Forsythe's presence. Now she knew that chill was warning her of this menace, and a shiver ran up her spine. She glanced at Sir Erick, who was actually gazing off in another direction. He didn't even seem to know Marcellus was in the room, much less standing right in front of Marissa. Queen Gertrude and Ruth also seemed oblivious to the old man's presence, which made Marissa even more anxious.

"You do not heed warnings, I see," he said, his voice low and menacing. "It will be a pity to see you completely broken."

"You will not break me, old man," she said, equally as menacing. "I did not make this trip only to be threatened by you. Whatever you have in mind, it won't work."

"Do not be so certain, child," he sneered. "As you can see, I have abilities. These abilities will take me far."

"Those abilities have not yet put you on the throne, so I put little stock in them. Now go away and stop trying to spoil my evening."

His eyes grew narrow in anger, and he took one step backward. "Do not underestimate me, child," he warned. "As of yet, you know nothing. Look around you. Does anyone else know I am here?"

She didn't need to look, she had already reached that conclusion.

"You see? I can shield myself and be well protected. Even you would not be able to discern my presence if I so choose."

She looked into his eyes as he made that claim, and saw a lie. *Could it be? Could she be able to sense him even when he could shield himself from everyone else? That would certainly be an advantage.*

"So," he continued, "I again offer you an easy way out. Simply return to the distant land from whence you came, and save yourself the humiliation and pain of being driven off. It will be so much easier for you."

"Sorry, Marcellus," she said evenly, "but I've never taken the easy way out. You might say, I love a challenge."

"More's the pity," he replied. "Well, you cannot say you were not given fair warning." And he took another step backward, raised his staff into the air, and vanished.

Marissa shook her head. "It seems," she muttered to herself, "that my real challenges are putting up with arrogant men who want to steal my throne."

Considering the tense beginning, the remainder of the evening went very smoothly. When the evening was over and she lay her head down on the satin pillow and pulled the satin coverlet over her shoulders, festive dreams filled her sleep. Merlin curled up on his own pillow next to hers, and purred her to sleep. All was well within the Kingdom of Avonridge.

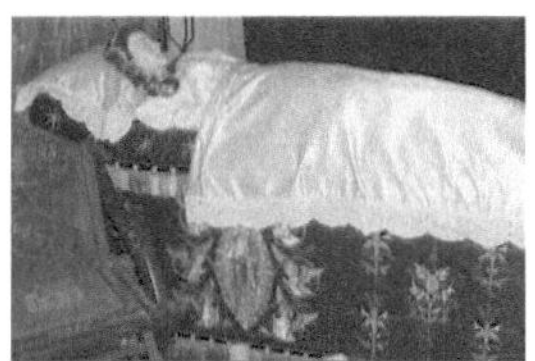

FOURTEEN

"The queen is dead! Long live the queen!"

Marissa woke with a start. Merlin was standing on the pillow beside her head, meowing plaintively.

"What?" she asked, still groggy. She sat up and heard the cry again.

"The queen is dead! Long live the queen!"

She rubbed her eyes and Merlin sidled up to her, rubbing his face on her arm. She reached out and scratched his head. Looking into his eyes, her heart was filled with dread.

"Is that true, Merlin?" she asked. "Is Gertrude dead?"

In answer, the cry came again, out in the corridor and coming nearer, "The queen is dead! Long live the queen."

Marissa grabbed the silk robe and hurried into it, then rushed across the room. She threw the heavy door open just as Ruth appeared.

"Oh, Milady," she whispered, tears welling up in her eyes, "hurry, you are needed."

Ruth turned and rushed down the corridor, and Marissa ran along behind, Merlin close at her heels. As they ran, shouts could be heard throughout the castle, shouts of "The queen is dead!" and Marissa's dread deepened.

Gertrude dead? How? Why? Marissa just arrived and needed Gertrude's guidance. And Gertrude had seemed in fine health yesterday. Why should she be dead now?

Ruth led her through the vast castle corridors until they reached a secluded, well-guarded wing. Armed guards lined the stone walls leading up to the huge wooden double doors. The doors were closed, and two guards stood before it, rigid as statues. Marissa could see tears glistening in the eyes of the guards.

As the two women approached, the guards, as one, grasped the doors' brass knobs and swung them open. Ruth stopped just outside the doorway and allowed Marissa to precede her into the room. Marissa hesitated, looked at Ruth's tear-stained face, then stepped slowly through the door, Merlin walking somberly beside her.

The dimly lit room held several candles against either wall allowing the only illumination. A huge, four-poster bed of carved blonde oak dominated the large room, with sideboards of matching wood on either side holding the candles. The same tall, narrow windows that vented the rest of the castle were here as well, but were covered with heavy white satin, keeping out the rising sun. White seemed to be Gertrude's color of choice, as the elegant satin coverlet that lay draped across the bed was also white.

And there on the four-poster bed, lying beneath the white satin coverlet, was the queen herself. Her long hair fanned out around her head on the pillow, giving the appearance of a soft brown halo. Gertrude's hands lay upon her abdomen, peacefully folded, and her eyes were closed as if still sleeping. Everything about the queen's appearance seemed to indicate simple slumber, but the deep chill within the room screamed of death. The chill was not physical, as the room was comfortably warm, but the atmosphere was certainly frigid.

Merlin hopped up onto the bed and stood next to Gertrude's shoulder. Marissa stepped closer to the side of the bed, her eyes never leaving the dead queen's face. She reached out a hand to stroke Gertrude's cheek, and was startled by the coldness of the flesh; obviously Gertrude had been dead several hours.

Marissa looked around the room, searching for anything that seemed out of place, anything that spoke of any form of foul play, but the room seemed peaceful as well. And deep within her heart, Marissa felt sure there was no cause to suspect any wrongful deeds. It almost felt as if she expected this, as if this was the natural course of things.

"So, it seems you have arrived just in time," the cold voice made Marissa spin to face the source. Merlin turned and arched his back, hissing softly. Standing in the doorway was the stooped, malevolent form of Marcellus. Marissa couldn't tell if he was deeply saddened, or completely joyous.

"You are not welcome here, old man," Marissa snapped. "This is a time for the queen's loyal subjects and those that love her, not a time for her adversaries."

"I assure you, dear lady," he sneered, "I was never an adversary of Gertrude. An adversary of the untried, green one that would assume Gertrude's place, however, is a different story."

"Majesty?" one of the guards said, and waited.

Realizing he was speaking to her, Marissa understood the guard was offering to get rid of Marcellus if she wished.

She shook her head, and said, "Thank you, but Marcellus was just leaving," the tone of her voice quite clear in its intent.

"As you wish, Majesty," Marcellus replied, the title nearly spat out of his cracked and brittle lips. "You are correct, now is not the time. Our time will come later," and on that, the bent and feeble-looking old man vanished.

The two guards and Ruth all gasped at the display of magic, but Marissa turned back to the still form of the dead queen.

"Gertrude," she said, barely above a whisper, "why didn't you tell me? There are so many questions I have, so many things I need to know. We could have talked last night, I could have learned so much in that short time. Now, what shall I do?"

"Your best," came a soft reply, and Marissa looked hard at Merlin. Again, he seemed to be speaking, again his eyes looked human.

"And you, little one," she said to the cat, "what is your place in all this? You've seemed to know so much right from the beginning.

Who are you, or rather, what are you?" But he just gazed back at her without a reply.

Ruth's soft sobbing brought Marissa back to the present. She moved to the grief-stricken young woman, and put her arms around Ruth's shoulders. Ruth gently laid her head against Marissa's shoulder, and wept quietly. Marissa could see the soldiers standing outside the doorway, and they too were weeping, their shoulders hunched and heaving slightly.

People had started to gather out in the corridor, looking into the queen's chambers expectantly. Marissa gave Ruth a last hug for comfort, and stepped to the doorway. All eyes were on her, waiting for her to address the crowd. She looked from one face to another, seeing grief on each one. The silence was broken only by an intermittent sob, and an occasional cry of, "The queen is dead."

Marissa took a deep breath, and said in a clear voice, "Yes, the queen is dead." Intending to continue, she was interrupted by loud sobbing and crying, and other cries of, "The queen is dead!"

She waited patiently for a few moments, then when the crowd quieted down sufficiently, she proceeded.

"Yes, the queen is dead," she said again. "Now we must honor her and attend to her in the proper manner. We will have a royal funeral for the whole kingdom to attend and pay their respects. We will send her to her eternal reward in the manner befitting the great queen she was."

Murmurs of consent went around the crowd. Marissa was about to ask the people to go back to their lives, when Ruth appeared beside her.

"Yes, dear people," she said loudly, "the queen is dead. Long live the queen."

"Long live the queen," was repeated around the corridor, and Marissa turned to Ruth questioningly.

"Long live the queen!" Ruth said again. Then turned to Marissa, dropped to her knees, and said, "Hail Marissa, Queen of Avonridge."

In unison, the crowd followed, dropping to their knees in the corridor. "Hail Marissa, Queen of Avonridge," echoed off the corridor walls. The two guards turned to face her, each dropped to one knee, and repeated, "Hail Marissa, Queen of Avonridge."

Marissa stepped back, overcome by the speed with which everything was happening. She had only arrived yesterday, and now she was expected to assume the duties and responsibilities of ruling the entire kingdom. Panic began to well up within her breast, when she felt Merlin rub against her leg. She looked down at him, and he looked up at her with warm, confident eyes. He meowed loudly, and the panic quickly receded. She drew comfort from his familiar face, the confidence he seemed to have in her. She reached down and scooped him up in her arms. He was purring loudly, and he rubbed his head against her cheek.

Looking out at the crowd of royal subjects in the corridor, her royal subjects, Marissa took a deep breath. Just a few days ago she was fending off the lascivious advances of her co-worker Rob, and dealing with unhappy customers on the phone all day. She had dreamed of a place where life was "challenging and exciting," and had certainly found just that place. Now, she felt overwhelmed, and wondered if she could live up to the challenge. Only Merlin's reassuring presence helped to calm her, and he continued to rub his face against her cheek, purring in her ear.

Since she had already decided to accept this fork in her road, she pushed down the fears and anxieties, gently placed Merlin on the floor, and again addressed the crowd.

"Good ladies and gentlemen," she said clearly, "I am grateful for your loyalty. As you must realize, there is much to be done. I must put the necessary events in motion. Please go back to your daily duties, so I may begin mine. I assure you, everything will be done with the utmost respect to the good Queen Gertrude. She will be laid to rest in the manner befitting her excellent reign."

Most of the women were still weeping, but all rose and slowly dispersed. The two guards rose, and resumed their places on either side of the doorway. Ruth stood and looked at Marissa questioningly.

Marissa gestured to Ruth to come closer, and she quickly obeyed.

"Yes, Majesty?" the young woman said.

"Ruth, I need you now more than ever," Marissa said. "I don't know anything about what to do. Obviously, we need to have a funeral, but I don't know how to go about setting it up. What do I do?"

Ruth wiped her eyes, and said, "Do not worry, Majesty. You need only to prepare yourself. Her Majesty, Queen Gertrude will be attended to by the proper persons. Word has been sent out to the whole kingdom, and this evening will be her funeral pyre. After that is completed, we will hold your coronation."

Coronation? Marissa again felt overwhelmed. She had never even witnessed a coronation, and didn't know what to expect. Living in the United States, she had only seen royalty from afar, and that was mostly through the gossip magazines.

Ruth seemed to misread Marissa's expression, and said, "Oh, Majesty, it will be done quickly enough. You will be crowned in a timely fashion."

Marissa looked at Ruth. She seemed to think Marissa was in a hurry to be done with the details so she could take her place on the throne. But nothing could be further from the truth. She was in no hurry to assume the responsibilities of ruling a kingdom.

Not in the mood to discuss her fears with the loyal young woman, Marissa simply nodded. She turned back to look at the old queen, a mixture of sadness, trepidation and anxiety filling her heart. Hearing noises out in the corridor, she turned to see several men hurrying toward the queen's chambers, carrying a litter made of blonde oak and white satin. It was roughly the size of a casket, approximately six feet long, three feet wide and two and a half feet high, and was supported by oaken poles that extended three feet beyond the confines of the satin box.

The men stopped in the doorway, and all bowed to Marissa. She looked at them, then at the dead queen, and nodded to the men. They walked into the room quietly, as if they were tiptoeing, and placed the litter on the floor next to the bed. One man opened the top of the litter, and it swung upward as if it were indeed a casket, then two men gently lifted the dead queen off the bed and laid her in the litter. Closing the lid, the men raised the litter and walked out of the room as Marissa watched in silence. Merlin stood at her feet and meowed, his voice full of sorrow and grief.

When the men were out of sight, Marissa said, "I better get back to my room to get ready."

"Oh no, Majesty," Ruth said. "These are your chambers now. I will have your belongings brought to you."

"My chambers?" Marissa asked. "Why?"

"Because these are the royal chambers, of course," Ruth said simply, and hurried out the door.

After she had gone, Marissa looked around the immense room. She spied a doorway she had not noticed before, a doorway in the far corner of the room. Peering through it, she saw another room, just as immense as the bedchamber, but arranged as a sitting room. She walked through the threshold and marveled at the elegance. It too was all in white satin, but instead of feeling cold, it gave the feeling of warmth and welcome. It had its own door into the corridor, so visitors didn't have to walk through the bedchamber to reach the sitting room.

A blonde oak table with two satin-cushioned chairs sat against the far wall, a chaise lounge upholstered in white satin against another wall, and various benches upholstered in white satin were scattered about. An oaken serving table stood tucked in a corner, and held two golden candelabra with white taper candles. Two tall windows were cut into the wall, but the white satin curtains were pulled back to allow the morning sun to light the room.

Ruth appeared in the doorway to the bedchamber, and said, "Majesty, we are ready for you."

Marissa obediently followed her into the room where a large bathtub had been placed, and a black satin gown lay on the dead queen's bed. No, not the dead queen's, the new queen's. The bedding had been changed, and fresh sheets adorned the mattress. There was no coverlet, and Ruth noticed Marissa's questioning glances.

"A new satin coverlet is being crafted for you, Majesty," she said. "The old one will be used to wrap the...," her voice faltered.

"I understand," Marissa said quickly. "This is going to be a difficult few days. You're going to have to help me with this, Ruth."

The young woman wiped a fresh tear from her cheek and nodded.

FIFTEEN

At dusk, the fire was lit on Gertrude's bier, and as the flames leapt into the sky, crying could be heard over the roar. Every man, woman and child of the kingdom mourned the loss of the great queen. During her reign, Gertrude had maintained peace with neighboring kingdoms. The dreaded Balvindor had terrorized the people, but less and less ferociously as time passed. It seemed even the fearsome dragon was tamed somewhat by the queen's gentle but firm rule.

The day had passed in a haze. Marissa was washed and primped and dressed, but it all felt like a dream. It seemed to pass in slow motion, yet night arrived before she realized. It seemed the whole kingdom had turned out for the funeral, and Ruth assured her it was indeed so, attendance being a requirement. Had anyone, man or woman, not attended, the other subjects would have punished them severely.

Even though the moon was full and stars filled the night sky, the mood was somber and beclouded. Soft weeping and the roar of the pyre resonated throughout the grounds. As the evening wore on, the sound of beating wings filled the air. All eyes turned toward the moonlit sky, and gasps erupted throughout the crowd. The dark form of Balvindor could be seen sweeping through the night. Marissa wondered if she had summoned the dragon by thinking of him.

The crowd was beginning to panic, and cries could be heard of, "The dragon! 'Tis the dragon come to massacre us! Save us from the dragon!"

Marissa stood up beside the flaming funeral pyre, and called out in a loud voice, "People of Avonridge, do not fear! Balvindor will not mar this solemn occasion." And she walked out among the crowd. The people parted and encircled Marissa, standing in a ring around her, giving her a thirty-foot radius. Standing alone in the middle of the circle, she lifted her face and voice to the dragon.

"Balvindor!" she called, "now is not the time for your childish antics. Please return to your lair and, if you feel it necessary to terrorize us, do so another day."

Balvindor swooped toward her, his huge form blocking out the moonlight. As he passed above her, Marissa could see that his talons carried something. He circled dramatically once more, then flew above the funeral pyre. Hovering over the flames, he seemed to be making sure all eyes were on him, as if the crowd's attention could actually be elsewhere. After pausing for several moments, he opened his huge talons and dropped what he had been holding. Again, the crowd gasped and drew back, but instead of anything malevolent being dropped, flowers rained down on the burning form of the dead queen. Many managed to adorn the ground around the pyre, but most floated directly into the flames, burning to ash before they touched the queen.

When the hail of flowers had finished, Balvindor swooped through the sky once more to fly above Marissa. "New queen!" he called. "You and I have met once, and we will meet again. Prove your right to follow the fallen queen, or die trying." And he flew out across the land, disappearing from sight.

Several hours later, the crackling of the funeral pyre could still be heard. Marissa could yet smell the odor of burning flesh, and gave an involuntary shudder. As Ruth helped her into a white satin nightgown, Marissa tried to replay the events of the day, but found many gaps in her memory. The only constant she could remember was the presence of Merlin at every moment. His little form was with her unfailingly, offering support and comfort, giving her the strength to last through the ordeal. The appearance of Balvindor was the only part of the day that was clear in her memory. When she had met Balvindor on the battlefield, he had spoken fondly of Gertrude. Now Marissa knew he had truly cared for the dead queen. Obviously, Gertrude had established her reign well, and Marissa knew she must do the same if she were to endure as Queen.

As she crawled into the huge four-poster bed, she found it funny, however, that she had not been allowed to sit on the royal throne, and instead had been seated on a carved, blonde oak chair beside the dead queen's bier. She had been given the queen's chambers before her coronation, yet she was not allowed to occupy the throne. Ruth said something about protocol and appearances, but Marissa hadn't paid much attention, it didn't really matter after all. All that was important now was a good night's sleep, as tomorrow would be just as trying and difficult as this day.

"Majesty, it is time to rise," Ruth's voice slowly penetrated Marissa's sleep-fogged mind, and she sat up, rubbing her eyes. She looked around the room in slight confusion, then remembered the events of the past few days. Yesterday they had sent Queen Gertrude to the gates of Heaven, the day before she had fought a dragon, and this day would bring her coronation.

It still seemed like a dream. Marissa had grown up in a democratic country with no place for kings and queens. She had believed that was the right way a country should be run, yet here she was preparing to become the monarch of a country she had never even heard of just a week ago.

"We must get you bathed and dressed," Ruth was saying as she went around the room taking clothing out of drawers and closets.

Merlin walked to Marissa and rubbed against her arm, purring loudly. She reached out and absently scratched his ear, watching Ruth fuss and hurry around gathering her clothes. Ruth carried a luminescent, white satin gown to the bed and reverently laid it on the mattress. She then returned to the closet and removed a long, royal purple cloak with gold embroidery around the edges. This she lay next to the gown, and rushed to the door to answer a knock.

Standing in the doorway were two men pulling the bathtub behind on a wheeled platform. They brought the tub into the room, and quickly disappeared. The rest of the preparations were all a blur to Marissa, as Ruth went about readying her for the day's events.

Marissa followed the procession led by Sir Erick down the corridor, still in somewhat of a daze. She vaguely remembered being bathed and dressed, and had to look down at her clothes to remind herself of the beautiful white satin gown she wore. She could feel the weight of the purple velvet cape draped over her shoulders, and marveled at its splendor. She knew the color purple signified royalty, and was glad she liked it.

As she marched down the corridor, the cape billowed out around her, and her hair danced freely about her head. Merlin trotted at her heel, and she felt comfort and reassurance at his presence, appreciating his unflinching loyalty. He was the one link she had to the world she had left behind, the world in which she had grown up. And yet, as each hour passed, it seemed he was so far removed from that old life, that he was as magical and curious as Balvindor himself.

The procession reached the main doorway, and guards swung open the massive oaken doors. Sunlight poured into the corridor, heralding a beautiful day.

Ruth echoed Marissa's thoughts, and said, smiling, "A fine day for a coronation."

Her presence also bolstered Marissa's spirits, and she smiled at the faithful young woman.

The procession walked through the doors and lined up on either side, allowing Marissa to step into the cool air of center stage. On each side of her stood six guards, dressed in royal purple, swords strapped to their sides, and holding ten-foot-long poles that each carried the kingdom's crest of the dragon. Flanking Marissa, the guards each made a quarter turn to frame her as she emerged onto the castle steps, and when she did, her heart skipped a beat and her breath caught in her throat.

Standing before the castle, as far as she could see, were thousands of people. Obviously, every royal subject turned out for the coronation of the new queen, just as they had for the funeral of the old. Marissa looked out at the throngs of people, and as they spied her, they began to cheer.

"Hail, Marissa!" thundered through the air, followed by two resounding foot stomps.

"Hail, Marissa!" was repeated, and two more foot stomps.

"Hail, Marissa!" came again, and yet two more foot stomps. Then the people began clapping their hands and yelling, "Yay! Yay!"

Marissa stood statue-still at the top of the stone steps, wondering if she would faint from the excitement, and total terror, of what she was about to undertake. Sensing her needs, Merlin rubbed against her leg, and meowed loudly enough for her to hear over the crowd. She looked down at the little black-and-white ball of fluff, who looked up at her with such human eyes, and her breathing grew easier and her heart slowed to a mere frantic pace.

As she looked at her faithful companion, he seemed to nod questioningly at her. She nodded back, and looked to Ruth. Ruth turned and raised her hand for silence, and the crowd immediately obeyed. She then stepped back out of view.

From the crowd at the bottom of the stairs emerged a bishop, dressed in a scarlet red cape and pointed miter hat. Old and stooped, he immediately made Marissa think of Marcellus. However, this man had an air of humility and respect, making her feel safe and welcome. He carried the gold and purple crown she had seen on Gertrude's head, and he walked to the base of the steps with it reverently held out before him, and stopped.

Silence hung in the air, and Marissa realized she was holding her breath. She forced herself to breathe, hearing each breath thunder through her head. Standing still and not knowing what else to do, she waited. The bishop raised the crown high, as if to let the crowd witness its beauty.

Looking into Marissa's eyes, the bishop called out, "This day we crown our queen!" His voice was clear, but slightly shaky, and he looked at Marissa with what seemed like a respectful challenge.

"Lady Marissa," he called. "It is your place to accept this crown and all the responsibilities that lie with it. You are called upon to become the ruler and protector of Avonridge. Are you capable of this calling?"

Having not been coached at all about this process, Marissa relied upon instinct, and replied in a strong voice, "I am!"

"Lady Marissa," the bishop continued, "you are called upon to become the ruler and protector of Avonridge. Do you desire this calling?"

"I do," she answered.

"Lady Marissa," he said, "you are called upon to become the ruler and protector of Avonridge. Do you accept this calling?"

"I accept," she stated firmly, and the crowd erupted in thunderous cheers. People clapped and danced around in circles, and the twelve guards pounded their flagstaffs on the stone, adding to the cacophony.

Marissa's heart felt like it would burst with joy, and she could feel immense love for each and every man, woman and child standing before her. These are my people, she thought, but then realized; no, not yet.

"Lady Marissa," the bishop called out, straining to be heard over the crowd. Those near him grew silent, and the rest of the crowd followed suit. Still holding the crown above his head, the bishop continued. "You have accepted the calling of your destiny. Come forward and be recognized."

As Marissa walked forward and began descending the steps, two men carrying a small litter made of purple velvet and oaken poles, came up to the bishop. He placed the crown on the purple velvet, and the men walked toward the crowd. The people stepped back, creating an aisle through which the men marched, proudly displaying the purple and gold symbol of the queen.

Marissa walked down the steps, the purple cape splayed out behind her like a royal fan. Merlin stayed tight at her heel, and the soldiers marched in an arc around her. Sir Erick fell into step behind the arc of soldiers, and Marissa could feel his comforting presence.

When she reached the bottom of the steps, the bishop turned and walked through the aisle behind the royal litter. Marissa followed, silently wondering where Ruth had disappeared to.

The crowds on either side stayed back a respectful distance, but every so often a hand reached out tentatively to try to touch the new queen. Not sure of the proper protocol, Marissa decided to be accessible to the people, and would gently touch their hands with her fingertips. Each time, she was rewarded with a, "Hail Marissa."

The bishop led her to a large wooden platform that had been erected in the spot where Marissa sat the day before. The chair she had occupied for Gertrude's funeral was removed and, in its place, stood the platform made of blond oak, purple velvet draped around its perimeter.

Stepping to the side of a set of steps, the bishop indicated Marissa was to ascend, and she readily obeyed. The procession of Royal Guards stopped at the steps, and the bishop followed Marissa onto the platform.

There was no place to sit, but a small X was marked on the center of the platform, so Marissa walked to it and stood, waiting. The bishop came up beside her and whispered, "Kneel, Milady," and she quickly did so. He then stepped to the edge of the platform, where the litter waited beside it, and lifted the crown.

The air hung heavy with silence, and Marissa again held her breath. As the bishop turned to face her with crown in hand, she looked past him at the clear morning sky. In the distance she could see a tiny speck that was steadily growing larger. She immediately recognized Balvindor sailing toward her, and stared at the approaching figure.

Seeing her expression, the bishop turned to see for himself what held her attention. When he spied the dragon, he visibly flinched, and in unison, the crown turned to see as well. Gasps erupted from the people, and they turned to Marissa.

As Balvindor drew closer, she could see the look in his eyes, and she was relieved. They held no malice, but she could see a challenge.

She rose to her feet and walked to the edge of the platform, facing the awesome beast.

"Balvindor!" she called, her voice echoing his challenge. "Come closer and be witness to the ascension of the new queen!"

The dragon drew nearer and stopped approximately thirty feet away from her. He called back in answer, "I shall be witness, but you still must prove yourself. When King Sorens is finished with you, if you still live, you and I will meet again. Until then, enjoy your ceremony."

"King Sorens?" Marissa repeated. "Who is he?"

"All in good time," was his only reply, and he hovered over the hushed crowd, waiting. Marissa returned to the X in the center of the platform, and knelt again. The bishop, an uneasy quiver on his lip, resumed the coronation. He walked behind Marissa and held the golden crown about four inches above her head.

"Lady Marissa," he said, only a slight tremor in his voice belying his false confidence, "you are called upon to become the ruler and protector of Avonridge, and you have accepted this calling. From this day forward, for as long as there is life within you, your only interest will be the welfare of your kingdom and your people. To this you will devote your thoughts, to this you will devote your energy, to this you will devote your life. There will be no other purpose in your life but the protection and betterment of Avonridge and her people. Swear this now before the people of Avonridge and before God, and accept your destiny."

Marissa heard the words, and instead of feeling overwhelmed or awed, she felt alive and exhilarated. She took a deep breath, and loudly stated, "I swear before the people of Avonridge and before God, that I will spend my life in service to my kingdom and my people. As long as I live, I will have no other purpose but to rule and protect Avonridge."

The bishop then placed the crown on her head and declared, "Rise Marissa, Queen of Avonridge."

The crowd erupted in renewed cheers, and Balvindor slowly drew closer. At the movement of the dragon, the crowd again became silent.

The beast flew straight toward Marissa, and she looked up at him and waited. She marveled at the realization that she felt no fear.

Balvindor flew about ten feet off the ground, slowly approaching the new queen. When he reached the edge of the platform, he began to rise. As his taloned feet soared over Marissa, close enough for her to touch them, one foot opened and a single red rose fell to the platform and landed directly in front of her. Balvindor then flew straight up, arched to the right, and flew back the way he had come. Within moments he became a mere speck in the sky, then disappeared completely.

Cheers once again resounded, and the people shouted in unison, "Hail Marissa, Hail Marissa!"

Marissa grasped the rose, then stood up. She raised the flower in silent salute to the beast that seemed more ally than enemy, then looked out at her people. *Yes*, she thought; *My people*. Her heart was so full she was afraid it might burst. For the first time in her life, she felt completely sure of herself. For the first time in her life, Marissa knew exactly where she belonged.

SIXTEEN

The celebration was to last a week. Pigs, lambs, steer, goats and pheasants all roasted on spits, filling the entire kingdom with delicious odors. It was late afternoon of the day of Marissa's coronation; she had been Queen for only a matter of hours, when Forsythe found her.

"Majesty," he breathed, and dropped to one knee, head bowed in what Marissa recognized as an exaggerated gesture. She had thought he reminded her of someone, but as yet couldn't recall whom.

"Congratulations, my queen," he said, looking up at her, still on one knee. His smile was sly and lascivious, and Marissa worked hard to resist the urge to slap him.

Instead, she quietly replied, "Thank you. I only hope I can do as fine a job as Gertrude."

"Ah yes, Gertrude," he said, and stood up. "Do not be fooled, Majesty. Gertrude was not the divinely extraordinary ruler these simple people want to believe she was. She had her faults."

"Don't we all?" Marissa replied coolly.

"Well, yes," he said, feigning ignorance of her contempt. "However, the queen is ordained by God. She should be able to make decisions with His wisdom."

"Ordained by God, yes," she stated. "But not God. We are still human. We ask His guidance, but we can only do our human best. If you expect me to be as wise and as strong as God Himself, I'm afraid I too will disappoint you." Her voice had grown icy and hard.

Forsythe seemed not to notice, and smiled, saying, "Oh no, Majesty. I do not for a minute believe you could disappoint me. After all, your coronation is but one element of your destiny. Becoming my wife will be another. I am sure God will make you realize that."

She raised her eyebrows at him. "You take liberties, Forsythe," she said, and the warning in her voice was unmistakable.

He raised his hands placatingly, smiling broadly. "Forgive my impulsiveness, Majesty," he said easily. "I will wait until you realize the truth of my words. I am forever your servant." He bowed deeply and sauntered off, hands clasped behind his green velvet waistcoat.

At that moment Marissa remembered whom he reminded her of; her coworker from her previous life, Rob. The lecherous smile, the arrogance, the belief he was God's gift to women, all echoed the man that had constantly hit on her at work.

She remembered the uneasy way Queen Gertrude and Ruth had both witnessed the initial exchange between Forsythe and herself at the banquet, and knew this would be a man to watch closely. His having stated he would marry Marissa clearly said he was after the throne. Marissa thought she would only have to contend with Marcellus on that score, but now she realized she was being naïve. She should have known attempts to steal her crown would be at every corner, and she must be ever vigilant.

She walked back to the throne, which had been brought out to sit in the courtyard, and dropped heavily into it. She still wore the white satin gown, purple cape and gold and purple crown, and was suddenly exhausted by the weight of it all. The physical heft of it was not taxing, but the emotional aspect had become a temporary burden. The sun began to set on this, the first day of her being queen, and she wished it would be over so she could crawl into bed and sleep for a week. Unfortunately, that would not be the case.

As Marissa sat quietly upon her throne, she looked out at all the people celebrating. Ruth was off to one side, talking and laughing with a small group of women. Sir Erick stood back against a stone pillar, keeping a casual eye upon the new queen. Marissa felt safe and loved, but wished desperately for the solitude and quiet of her chambers.

While silently longing for her bed, a young man approached her, tentatively bowing and looking questioningly at her.

Marissa sat up straight and motioned the young man forward, saying, "Come, sir. What's on your mind?"

The young man stepped closer, appearing quite uneasy, if not completely frightened. At the sight of the anxious young man, Sir Erick approached and stood beside the throne, as always giving Marissa his complete and undying support.

"Queen Marissa," the young man started, his voice croaking like a frog's. "I have a message from King Sorens of Southcross." He silently waited for her response.

Marissa had only heard of King Sorens from Balvindor this very morning, but now knew he would not be an ally. By the fear this messenger was exhibiting, it was clear he did not bring the new queen congratulations. She motioned for him to continue.

"Queen Marissa," he said again. "My lord, King Sorens, is issuing you an ultimatum." The young man's voice cracked again, but he continued. "You are to turn over rule of Avonridge to him by sunset tomorrow, or face the king and his soldiers in battle."

The crowd around Marissa had grown silent at the mention of King Sorens, and now gasps could be heard all around. Sir Erick sucked in his breath, and Marissa could feel the anger burning in him.

The setting sun cast long shadows across the courtyard, and silence hung in the air like a heavy fog. Everyone waited to hear Marissa's reply to this ultimate challenge.

Looking at the young man trembling before her, Marissa thought, what a coward Sorens must be to send a boy to do what should be a knight's job at the very least. Obviously not a soldier in any respect, but seemed to be nothing more than an expendable pawn, he looked terrified at what she might to do in answer to this insult. He could not even bring himself to meet Marissa's gaze.

"What is your name?" she asked in a strong voice.

He gulped and looked up through his eyelashes at her. "Um, um," he stammered.

"What is your name?" she asked again, but softened her voice somewhat.

"Um, I am Callus," he croaked, and continued to tremble so violently Marissa wondered if he might fall over.

"Callus," she said calmly, "you have no reason to fear me, I do not punish the messenger."

He looked skeptical, but nodded obediently.

"Callus, you may bring my reply back to your lord. As generous an offer as the former is, I am afraid I must choose the latter. If King Sorens wishes to rule MY kingdom," and her voice came down heavily on the word 'my', "he must take it for himself. Please advise your lord that I will be waiting to meet him on the battlefield at sunrise two days from now."

Callus nodded and turned to go, but Marissa called him back.

"And Callus," she said, and he faced her again. "Please also advise your lord that I do not take kindly to insults. I will remember the fact that he could not send a messenger of a more worthy station to face a queen."

Marissa hadn't thought it possible, but Callus paled even more.

"Tell your lord that, not only will I defend my people, I will also defend my own honor."

Callus nodded again and stepped back. Just then a man in the crowd shouted, "Kill the one who challenges Queen Marissa!" and concurring shouts went around the crowd.

Callus shrunk down, looking like he wanted to crawl into a hole somewhere.

Marissa stood up quickly and raised her arms for silence. The people obeyed, but begrudgingly.

"People of Avonridge," she said, "I agree, we will fight the one who challenges me! But it is not this man. This man is simply the messenger of a coward."

Callus looked at Marissa with a mixture of fear and anger, but said nothing.

"This man," she continued, "will be guaranteed a safe passage. He will not be held responsible for the actions of his king. That would be repaying cowardice with cowardice. This man is to go back to his king and deliver my reply, and he will do so in complete safety." There was no question in Marissa's voice, she expected full compliance, and would have it.

Callus hurried through the crowd unmolested, however slurs were thrown at the hapless young man as he scampered away.

Marissa turned to the Black Knight. "Sir Erick," she said quietly. "Who is my Captain of the Guard?"

"That would be Roland, Majesty," he answered.

"Bring him to my chambers," she said, then turned and strode to the palace.

Ruth was in Marissa's chambers when she arrived, already setting out the golden armor. When Marissa entered, the young woman dropped a quick curtsey, and hurried to retrieve the purple cape and golden crown from the new queen.

As Ruth helped her out of the white satin gown, Marissa said wistfully, "This is all happening so fast, Ruth. I haven't been given the chance to catch my breath since I first met Sir Erick."

"That is the way of things, Majesty," she answered.

"And it seems like a lifetime ago," Marissa continued. "The day I met Sir Erick, and you, it all feels like years have passed but it's only been a matter of days. I thought my challenge was to face Balvindor, but it's been one thing after another."

"Your first challenge was to face Balvindor," Ruth said. "That was but one of many." She helped Marissa step into the golden suit of armor, fastening the breastplate into place.

"And you met that challenge admirably," Ruth said. "I know you will continue to meet your challenges equally as well."

"I hope so," Marissa said. "But this is all happening too fast. I've been Queen for hours, and now I have to lead an army into battle. Am I ready for this? Thousands of men's lives will all depend on me and my leadership. How do I know I'm ready?"

"You will know, Majesty," Ruth replied with confidence. "You will see, you were born for this. You will rise to the challenge, just as you did facing Balvindor."

"Well, I'm sure Sorens thinks he can intimidate me. After all, I haven't even been Queen for a full day yet. And how does he feel about women in general?" she asked.

"I am not sure I know what you mean, Majesty," Ruth said, avoiding Marissa's eyes.

"Ruth," Marissa said, "you know exactly what I mean. Forsythe has already let me know he thinks a woman can't do this job properly. Do most men feel that way here?"

Ruth refused to meet Marissa's eye for several seconds, then relented and nodded. "Yes, Majesty. Not all, but some men in this kingdom feel women should be subservient. Fortunately, most that feel that way are intelligent enough to keep their feelings quiet. Unfortunately, King Sorens feels no need to keep his feelings to himself. Being a king, there are not too many who would challenge him."

"Well, he's about to meet one who will challenge him. He will be quite surprised if he thinks I'll just quietly allow him to take my kingdom. I'm used to chauvinism, and I can deal with it. Any man that discounts me simply because I'm female has quite a shock coming to him."

Marissa then slid the gold chain mail headdress over her head and smoothed the skirt around her shoulders. As the last piece of armor was fitted into place, she could feel it come alive. Warmth and life raced through the gold, whispering to her of strength and ability. Through images and thoughts, the armor reassured her and bolstered her confidence.

She enjoyed the feeling of oneness with the armor, reveling in its connection with her spirit. She began to feel the room grow dimmer, and Ruth faded from view. Taking slow, deep breaths, Marissa welcomed the messages from the armor, and it spoke to her, conveying knowledge and wisdom through images and feelings.

Through a fog-like vision, the armor told Marissa what the coming days would hold. It revealed to her the ensuing battle with Sorens and his army. It showed her the horror of warfare, the destruction of human beings, bloodshed and anguish. She stood statue-still as image after image flooded her mind, showing her what this battle would do to her kingdom, and ultimately who would prevail. Marissa saw men lying in destruction, horses torn apart in battle.

As the fog lifted and the room came back into view, Ruth stepped forward and grasped Marissa's arms.

"Majesty," she called softly, concern in her eyes. "Are you well?"

Marissa took a deep breath, not feeling at all dazed, and replied, "Yes, everything is going to be fine."

Ruth nodded, relief on her face, and said, "Good. Sir Erick and Roland await your presence."

Marissa walked through the door into her sitting room, and Sir Erick and the Captain of the Guard turned to face her. Both men bowed deeply at the waist, and waited for instructions.

Walking to the table across the room, Marissa said, "Come gentlemen, we have much to discuss."

SEVENTEEN

The autumn morning dawned bright and crisp. The cool air hinted at a beautiful fall day. Twenty-four hours before she was scheduled to face king Sorens and his army, Marissa was ready. The night of her coronation had been spent in preparation for the coming battle, and neither she, Sir Erick nor Roland had been able to sleep a wink. They had spent the whole night in her chambers; planning, sending messengers to ready her army and gather supplies, until they could hear a rooster crow, heralding the breaking dawn. The trio had decided their best chance would be to arrive on the battlefield before Sorens. They would set up camp, survey the area, and find the best location to give them an edge in the battle. Roland's intelligence had already told them they would be outnumbered by thirty percent by Sorens' army, and they had to even the odds somehow.

Breath came out as steam in the cool morning air. Marissa sat atop a pure white mare, purple reins in hand. Dressed in the golden armor, she wore the royal purple cape, which draped out behind her, covering the mare's rump and hanging down the sides to just inches above the ground. The Dragon Slayer was strapped to her waist and hung down beside her leg. Half a length behind and to either side, rode Sir Erick and Roland. Of course, the Black Knight rode a black stallion, and Roland's steed was a rich bay color.

Marissa rode across the drawbridge, the two men closely flanking her. The horses' hooves made sharp sounds as they marched across the huge wooden planks. When they reached the outside of the castle courtyard, the very spot where she had faced Balvindor for the first time,

Marissa stared at the sight. Standing before her, as far as she could see, was an immense army. Men dressed in heavy leather tunics, wearing swords and carrying spears, sat atop horses draped in protective leather. Standing in formation, ready to obey their leader's command, her army was magnificent to behold.

Roland had informed her there were five thousand mounted soldiers, but looking at them now, it seemed like ten times that many.

Off to the side were the foot soldiers, those that did not ride. These amounted to thirty thousand men, all with swords strapped to their waist, and shields on their arm.

Behind them were the archers, ten thousand strong, also on foot, bows in hand and quivers filled with arrows strapped to their backs.

Marissa's head felt foggy, and her breathing grew rapid as she surveyed her army. Every man before her had placed his life and loyalty at her feet, and it was her duty to keep them safe. The cool air filled her lungs, calming her racing heart, and she silently vowed to protect not only these fighting men to the best of her ability, but each man, woman and child for whom these men were fighting. Sorens would not find her an easy mark. If he wanted to take all of this away from her, he would have to work, and work hard.

She moved to the front of the army, every eye on her as she rode to take her place. Facing the men that would put their lives on the line for her and her kingdom, she again found it hard to breathe. She forced herself to become calm, knowing it would be difficult, if not impossible, to think clearly if she let her emotions run unchecked.

She took a deep breath, and addressed the army.

"Soldiers of Avonridge," she called. "Today is the day you have trained for. Our kingdom is being threatened, and with it our way of life. Avonridge is a peace-loving kingdom. But if we must fight to protect her, then we will fight!"

Cheers went around the army, as each man shouted his agreement. Marissa waited for them to become quiet, then continued.

"I will not allow anyone to threaten Avonridge. I made a vow yesterday to defend her, and I will keep that vow. As long as there is breath in me, I will stand against any and all enemies. And you, my faithful soldiers, will stand with me. We will present a united front, and we will prevail!"

Again, cheers echoed through the morning air, and swords were raised in salute. "Hail Marissa! Hail Marissa!" resounded through the clearing. She then turned her mare and, with Sir Erick and Roland beside her, moved her army forward.

"Majesty," the scout called, breathing hard. "Majesty, the enemy is still a half day away."

Marissa, Roland and the Black Knight stood outside the tent that served as headquarters, warming their hands on mugs of hot mulled cider. A small fire blazed at their feet, giving warmth and light to the late evening darkness. The queen greeted the scout.

"Good, Larson," she replied. "Come, warm yourself."

He jumped from his horse's back and walked to the fire, respectfully bowing his head. He extended his hands to the fire, and Marissa waited patiently for him to speak.

Roland, however, was not so patient, and growled, "Your report, Larson."

"Yes sir," the man said quickly. "King Sorens and his army are marching this way. I estimate he has eight thousand mounted, but the foot soldiers number a little less than ours. His archers may outnumber ours by only a thousand or so."

"And you said they are a half day away?" Roland said.

"Yes sir, by my calculations about five or six hours. He is pushing them, and I think he's trying to beat us here." Larson smiled at that, enjoying the slight edge Queen Marissa held over King Sorens.

"Good," Marissa said. "We will be ready. Roland, do you think Sorens knows the terrain as well as you do?"

"No, Majesty. This is still Avonridge land. He would not have explored it as thoroughly as his own kingdom's."

"Still, I can't assume he's completely unaware," she said. "I must be prepared for anything he has up his sleeve. It will be dawn in about six hours, and Sorens should be here by then. Send word to the men they have exactly three hours to sleep, and I strongly suggest they use it."

Roland turned to a guard standing beside the tent and barked the order. The guard immediately hurried away. Marissa then turned to the scout.

"Larson, you've done well. Now you get some rest also. We'll need every man when we meet the enemy."

"Yes, Majesty," he said, then bowed and hurried away.

The guard returned to his post beside the tent, and stood at attention.

"Let's go inside and put the finishing touches on our strategy," Marissa said, and she entered the tent, followed by the Black Knight and the Captain of the Guard.

One hour before dawn Queen Marissa sat astride her white mare, Sir Erick to her right and Roland to her left. She wore the gleaming golden armor, and the purple velvet cape was draped around her shoulders. She gazed out across a great expanse of field grass, muted and smoky in the predawn darkness. She could not see King Sorens and his army, but could hear them as they settled their camp.

Stretched out behind her sat her mounted army, spanning hundreds of yards to either side. Behind them stood the archers, bows in hand. The foot soldiers were grouped to the sides, swords ready. Marissa would save them for after the initial engagement.

The day grew steadily brighter, and she could see shadows of her enemy across the vast field. They were not in any battle formation, still making camp after the march that had lasted all night.

As the first ray of sunlight flashed across the field, Marissa called out, "Sorens! I am waiting for you."

It took several seconds, but a reply came back, gruff and angry. "Marissa, the battle will not begin until I am ready."

"And you are not ready?" she taunted.

"Tomorrow is the day we shall meet," came the reply. "That was my offer."

"Your offer means nothing to me," she called. "Shall I go back to my people and tell them you are afraid to face me?"

"I am afraid of nothing!" he bellowed, and Marissa knew she had hit a nerve. "You will abide by the rules of fair engagement," he demanded. "Tomorrow we shall meet."

"Tomorrow you shall play this game alone! I will take my army home and celebrate the cowardice of a king that makes a threat, but cannot follow through. Really Sorens, you disappoint me! Men," she called in mock disgust, "we return to Avonridge and forget this little flea of a man."

The sun was completely over the horizon now, rising behind Marissa and her army, bright and fiery. She could see the army across the field. A heavyset man with a long gray beard and long gray hair was just mounting his horse. The queen was heartened even more by the fact that Sorens and his army would be advancing directly into the sun.

"You will regret calling me a coward, Marissa!" he roared. Behind him, men on horseback were lining up. Even from this distance Marissa could see the weariness in their manner, and a pang of sympathy for them struck her heart, but she quickly pushed it aside, and rejoiced that her ploy had worked. Not allowing Sorens' army to rest would even out the playing field, since he had over fifty percent more mounted soldiers than she.

"Well done, Majesty," Sir Erick said softly.

"Now we wait," she replied.

Just as Roland had said, Sorens was an arrogant man. Being called a coward was bad enough, but being called a coward by a woman must be excruciating. Sorens had risen to the bait and was preparing to lead his army into battle, even though they had marched for more than twelve hours with no chance to rest.

"I gave you the opportunity to relinquish Avonridge without bloodshed," Sorens called. "You refused and now you will regret your foolhardiness!" He stretched out his arms in a dramatic flourish, and shouted, "To the enemy!" spurring his horse to a full gallop.

The men behind their king echoed his cry, and rushed forward, galloping hard, swords drawn.

"Archers!" Marissa called, and arrows flew into the air at the onrushing assault. When they landed, screams could be heard, and several horses crashed to the ground.

"Archers!" she called again, and a second volley of arrows flew, with results just as accurate and devastating.

The ground shook at the approach of the enemy, and Marissa's heart pounded in her ears. Her mare snorted and stamped a foot, and she petted her neck to quiet her.

The tension in the air was great as the army of Avonridge waited for Sorens and his men to close the distance. Horses fussed and stamped, tossing their heads. Having been trained for battle, the horses were eager to get started. But still Marissa waited.

She could see Sorens' gray hair flying behind him as he raced toward her. She could see arrogance on his face as he was sure victory would be his. Then she could see surprise as his horse went down beneath him and he was thrown forward onto the ground. All around him horses went down, throwing their riders. But instead of landing on hard ground, the field grass seemed to swallow them up. More horses came, more horses fell.

Before they could comprehend the situation, the onrushing masses were trampling their fallen comrades, and being thrown down to be trampled themselves. Marissa could see confusion and panic on the faces of the exhausted soldiers as the ground seemed to swallow up not only them but their mounts as well. As each man tried to rise, they were bogged down and trampled by the soldiers behind them. Horses screamed in frustration, as they struggled to but couldn't rise. Each wave of oncoming riders made slightly farther progress, as they unwittingly rode right over their downed comrades and horses. Blood helped turn the scene a bright red, as horses' hooves slashed through the flesh of man and beast.

Marissa had chosen the battleground well. The earth between her camp and Sorens', the very ground they were now trying to cross, was not solid, but a marshy swamp. Vast bogs of mud were not only impeding the intended onslaught, but were sucking the soldiers and their mounts down, making Marissa wonder if in fact quicksand filled the swamp.

Shouts and cries of terror filled the air, as Marissa and her army watched the enemy flail around, trying to free themselves. The morning continued to brighten, marking a stark contrast to the darkness of the hideous scene before her. Horses tried to rise, throwing their heads about in an effort to gain some foothold, but sinking deeper into mud and mire. Men grasped their mounts' necks, trying to pull themselves out of the suffocating swamp, only to be dragged down further.

As many as two thousand men and their horses were trapped in the mud before those behind them could stop their charge. Confused and daunted, they drew back toward their camp, waiting for further commands from their leader, who was himself trying to break free of the stifling marsh.

Marissa watched Sorens climb onto his horse's neck, trying to gain some ground. He managed to break himself free of the mud, and with so many of his soldiers and their mounts close by, he climbed from one fallen animal to the next, slowly working his way back toward his camp. As Marissa watched, the king used his own men as stepping-stones, walking on their half-sunken bodies, pushing them down further in his effort to return to the safety of his camp. Slowly he made his way, sacrificing man and horse to save himself.

Torn between compassion for Sorens' soldiers and duty to Avonridge, Marissa knew she must be ruthless.

"Archers!" she shouted again, and again a volley of arrows flew through the air. Again, they hit their mark, and men and horses screamed anew in pain and death. However, Sorens was not hit. Marissa wondered at his good fortune. He managed to make progress back to his camp. Every now and again, he would lose his footing and sink into the mud, only to have a faithful soldier help him out to continue his escape. Each time a soldier helped the king break free of the mud, the king would reward him by using him as a foothold, stepping on whatever was available; shoulder, back, head.

Marissa shook her head at the self-serving man that obviously cared nothing for the men who willingly gave their lives in service to him. It broke her heart to watch this needless sacrifice. Sorens' arrogance and stupidity were killing his soldiers in a futile attack. As she watched, she could see the king make his way to the safety of his camp, and break free of the marsh. With him, only a handful of soldiers made it to solid ground, and they were covered in mud and stooped with fatigue and defeat. Perhaps half a dozen horses were lucky enough to free themselves of the mud, and ran in panicked frenzy, not allowing anyone to catch them.

Marissa waited a few moments, listening to the shouts of the enemy soldiers, as they tried to regroup. When Sorens finally managed to calm his army, she called out to him.

"So, king," she shouted, "have you changed your mind? I still await your attack."

"Marissa!" he roared in anger, "you will regret this!"

"I regret only that we still have not met," she taunted. "I am prepared for battle, yet you seem to only want to stroll through the fields. My soldiers are anxious to defend their queen, pity they don't have a worthy opponent."

"You will pay for your trickery!" he shouted. "You disregard the rules of fair play, and that will not go unpunished!"

"Fair play?" she called. "I should allow a usurper to take my throne because of rules? Really, Sorens, you can do better than that."

While she was engaged in conversation, Roland led half the mounted soldiers off to one side, while Sir Erick took the rest to the other. Flanking the marsh, the archers were standing prepared, arrows notched in bows, and the foot soldiers were fanning out behind their queen.

"I will do better, Marissa!" the king answered. "By the end of this day, your head will adorn my castle gates. You will regret ever having returned to Avonridge."

Sorens had acquired a new horse, and was mounted and gathering his soldiers, when from either side the army of Avonridge rushed together, swords drawn, converging upon the exhausted enemy. Sorens turned to see Sir Erick leading a mounted assault from one side, then spun around to see Roland charging from the other. As he turned back to face Marissa, his eyes widened in surprise and horror, as the queen in the golden armor, mounted on the pure white mare, came charging across the very bog that had engulfed his men.

Having had time to completely survey the area, Marissa had been given a detailed map of the marsh, and knew of a three-foot wide strip of solid ground. She now used this tactic to instill even greater fear and awe in her enemy, and it seemed to be working spectacularly, as Sorens' face grew white at the sight of the queen's advance, over what had just been uncrossable ground. Marissa held the Dragon Slayer high as she sped across the field of carnage, her purple cape flying out behind her like a banner of conquest.

Sorens leaned back, flinching at the ferocity of the unexpected attack, and his horse reared up in protest at the yank on the reins. He managed to hold his seat, but was unsettled enough to add to his agitation. Marissa knew the best way to fight a formidable opponent was to keep him off guard, and she threw everything she could at him. His anger and frustration were evident on his grizzled face, as his mounted soldiers were being overpowered by Marissa's army.

Taking full advantage of her enemy's fatigue, Marissa charged onward, screaming out her battle cry of, "Avonridge! Avonridge!"

As sword met sword in the crisp morning air, the resounding clang of metal on metal filled the senses, and her soldiers echoed her cry. "Avonridge! Avonridge!" rang through the air as Sorens' army tried to draw back from the onslaught.

Sorens gathered his wits enough to concentrate on the queen bearing down on him, and raised his sword in reply. Barely slowing her charging mare, Marissa slashed the Dragon Slayer at the man that would steal her kingdom. Her sword met his, and the force of the blow reverberated up her arm. She was relieved, however, to see Sorens jolted almost from his horse. He leaned back, grasping desperately at his horse's mane to keep from falling to the ground, and dropped his sword in the process. Marissa spun her mare and charged again, bringing the Dragon Slayer across in front of her, catching Sorens across his armored chest, and knocking him to the ground. She then jumped from her horse's back and prepared to meet the king in hand-to-hand combat.

Marissa's foot soldiers rushed in to join the battle, swords raised to defend their queen and her kingdom. Shouts of anger, cries of pain and the clanging of metal against metal filled the air. All around, the sounds of battle echoed through her ears, but grew steadily dimmer, as all her concentration centered on bringing down the leader of this offense.

Sorens quickly retrieved his sword, and stood to face the enraged young woman that charged to meet him. His eyes were filled with emotions; surprise, anger, and so many others. But he seemed to grow more assured when he could see just how tiny Marissa was compared to his bulk. He even let out a laugh at the sight of the petite young woman wielding a sword that seemed as tall as she.

"Are you sure you can manage such a large weapon, Marissa?" he asked with mock interest.

She stopped her advance just two yards or so away from him, and smiled in return. "Why, Sorens, you almost sound concerned for my welfare. Could it be you really have a soul after all?"

He stopped smiling when the realization of the insult hit him, and he flared out in anger, slashing his sword at her in a menacing arc. She blocked with the Dragon Slayer, and quickly spun away, bringing her sword around in a low circle. She caught Sorens' leg just below the knee and knocked his feet out from under him. He landed with a crash of armor and rolled away, surprisingly agile for a man of his age and size.

He tried to rise, but Marissa moved quickly and, just as he reached his knees, she attacked again. The Dragon Slayer slashed through the air in a downward movement, and almost hit its mark, but Sorens was just quick enough to block it before it sliced into his neck. He was knocked back slightly, and leaned at a precarious angle, as Marissa again slashed at him. Again, he managed to bring his sword up to block the blow, but this time he quickly swung it down and across and knocked Marissa onto her back as well.

She rolled out of the way and quickly gained her feet, but Sorens was up as well, and came at her, slashing and swinging his sword. She blocked blow after blow, being driven backward. Suddenly she sidestepped just as Sorens was bringing his sword down, and, expecting resistance and finding none, he lost his balance. He stepped heavily forward and Marissa swung the Dragon Slayer across and caught him on the shoulder, knocking him facedown onto the ground. He rolled to try to evade her attack, but the fatigue of the all-night march had taken its toll, and he was too slow. Just as he turned over, his eyes locked on Marissa's, and the Dragon Slayer was pointed directly at his throat just above his armored chest plate. Anger, hatred, and many other emotions blazed in his tired eyes as he looked at the young woman who had bested him.

The battle that raged around the two leaders slowly came back into focus, as the combatants realized the balance of power was unevenly matched. Sorens' men were being over-powered at every turn, and when they realized their king had fallen, they began to panic. Some threw down their swords, others simply turned and ran. One soldier, however, came running toward Marissa and his king, sword held high in attack. He struck at Marissa, and she deflected his sword, but in so doing, Sorens was able to rise to his feet. Sir Erick appeared out of the melee, and drew the soldier's attention from Marissa.

As Sir Erick and the attacker fought, Marissa turned to face Sorens, sword at the ready. But instead of resuming the fight, Sorens fled back toward his camp. One of his soldiers was nearby and still mounted, and Sorens ran to him, grabbed him by the arm and threw him to the ground. The king then jumped onto the horse's back, and loudly called to his army.

"Retreat!" he shouted. "Retreat!" And he spun the horse and raced away. His soldiers that were engaged in combat faltered, not knowing exactly what to do. Some managed to follow their fleeing leader, but others were unable to, still battling the defending army. Again, Marissa felt sympathy for her enemy's soldiers.

"Men of Avonridge!" she called. "Stand down." Immediately the sounds of metal on metal ceased, and the fighting stopped. Marissa's army stood surrounding Sorens' soldiers. She caught her mare's reins, and climbed into the saddle.

"Soldiers of Southcross," Marissa shouted. "You are defeated by Avonridge, and your king has abandoned you! Drop your weapons and you will live. Fight me, and you will die!"

Realizing the futility of any further resistance, those soldiers that hadn't yet dropped their swords, immediately did so.

Marissa rode through the crowds of men, surveying the battle scene. Soldiers lay dead and dying on the ground, most of them the enemy. She was relieved to see very few of her own soldiers among the casualties. Off in the distance she could hear the sounds of retreating soldiers, those that were lucky enough to have broken free and were fleeing along with their king.

"Round up the defeated soldiers!" Marissa called. "They will be our guests at Avonridge." She then turned her mare and rode back across the marshy ground to her own camp, knowing the details would be well taken care of by Sir Erick and Roland.

EIGHTEEN

Cheers and shouts of "Hail, Marissa! Hail, Marissa!" resounded through the crowd as the army of Avonridge returned. She, Sir Erick and Roland led the procession through the gates into the castle courtyard, followed by the archers and foot soldiers. The mounted army was herding the captured soldiers, which amounted to about two thousand men. Behind them the spoils of war; the horses they managed to round up and claim for Avonridge. Marissa had faced another challenge, and again emerged the victor.

Ruth awaited her queen to greet her with a wreath of roses to hang around Marissa's neck. Marissa rode to the small platform where Ruth stood, and leaned down for the young woman to place the wreath over her head. With this accomplished, the crowd increased their cheering.

Roland saw to the encampment of the prisoners, while Marissa went to the castle steps. Dismounting, a page quickly led her mare to be cared for, and Marissa climbed the stairs. When she reached the top, she stood facing her subjects. The courtyard was filled to overflowing with cheering men, women and children, and Marissa again felt awed by the love of her people. It astonished her that she would be accepted so completely, so unquestioningly. She had only come to Avonridge a few days ago, but it felt like she had lived here all her life. Thinking back, it became more difficult with the passing of each hour to remember her life before that fateful day when she met Sir Erick.

As she stood before the crowd, her golden armor gleaming in the last rays of the setting sun, she looked out over the people. Something caught her eye, and she peered into the sky and spied a familiar dot. As the crowd caught sight of Balvindor sailing toward them, they gradually became silent. The dragon approached and, when he drew to approximately thirty feet from Marissa, hovered over the people, gently flapping his delicate wings.

Silence was thick as Balvindor gazed at the queen in the golden armor standing atop the castle steps. Finally, the dragon said, "You have met another challenge, Majesty." He then made a sideways movement with his head, and flew several feet to his left, and resumed hovering.

Marissa watched him in silence, waiting to see exactly what he would do. Balvindor continued to gaze at her, then said, "I am impressed with your ability. First you were able to stand against me admirably, and now you have defeated King Sorens. Perhaps good Queen Gertrude did, in fact, make the correct choice."

"Thank you, Balvindor," Marissa said. "I have only done what was needed."

"Indeed," he said, the gentle flapping of his wings just barely audible over the hushed crowd. "You show great promise, Majesty. These were but two of your challenges, yet you have done well. It remains to be seen how well you do against what is yet to come. But I have decided to accept you as sovereign of Avonridge." He moved his head again and flew back to his right, hovering in his original spot.

"I wasn't aware you had any choice," Marissa replied, "but I appreciate your sentiment."

"Do not underestimate my influence, Majesty," he warned. "Keep in mind that a sovereign who cannot protect her people from the local dragon does not remain sovereign very long."

"Point taken," she agreed. "Then, as a symbol of our truce, may I offer you some refreshment before your long journey home?"

Balvindor lazily flew from side to side for a few moments, considering her offer, then replied, "Yes, I believe I shall partake of your celebration." He flew to a nearby fire, over which a large spit with a whole slaughtered steer rotated, reached out with taloned claws and grasped the steer. He lifted it easily, the spit still attached, and flew back in the direction from which he had come.

Marissa watched him go, gently shaking her head. "Strange beast," she muttered quietly, then turned her attention back to the crowd before her.

"People of Avonridge," she called, "today we have been victorious!" Cheers resounded throughout the courtyard. She waited for them to quiet down, then continued.

"Today we have defeated an immense threat to our peaceful kingdom. We cannot believe, however, that it is over. Today we faced King Sorens."

At the mention of the king's name, the crowd erupted in boos. Again, Marissa waited patiently. When they quieted, she went on.

"Yes, we defeated King Sorens and his army, but we cannot, for a moment, think he will accept that graciously. He has not only been defeated; he has been humiliated. And we must be prepared for his retaliation."

The crowd grew grim as they listened to their queen. She didn't want to hamper their celebration, but wanted them to be alert for any signs of danger that might sneak in.

"King Sorens will regather his army and strike again. That will take him some time. The celebration for my coronation was so rudely interrupted, and he was made to pay for that intrusion." Again, the crowd cheered.

"But now we have something else to celebrate," Marissa said. "We can have a double celebration; one for my coronation, and one for the defeat of King Sorens. And celebrate we will!" Cheers and shouts of "Hail Marissa!" echoed off the stone walls of the castle.

"People of Avonridge!" she called, and they quieted. "We will rejoice in the victory of our glorious army, but we will keep an ever-vigilant eye open for any treachery. King Sorens was straightforward in his first challenge. I doubt he will be so again. I am expecting a more devious encounter from him. Avonridge needs every one of her subjects to be alert for any activity. The survival of Avonridge depends on the loyalty of her subjects. You good people are our eyes and ears, and you will ensure the safety of our kingdom. But for now, we celebrate!"

Cheers erupted again, people danced and clapped their hands, and the celebration resumed as if uninterrupted, except that it grew in magnitude as the army's victory was included.

The captured soldiers were housed in an unused part of the stables, and given food and water. They seemed to welcome the quiet, and many fell asleep almost immediately. After being pushed by Sorens to the battlefield, given no chance to rest before the battle, and then marched back to Avonridge, they seemed happy to just lay down and take advantage of Marissa's hospitality.

Roland had originally taken up a post guarding the prisoners, but Marissa insisted he join in the celebration. After all, he was instrumental in the victory, and she wanted him to take part in the reward.

She had taken a quick bath and was once again wearing a glorious satin gown, only this one was a deep ruby red. She wore the gold crown, but wished she could have left it in her chambers. It wasn't particularly heavy, but she thought it ostentatious. Ruth, however, reminded her she was now Queen, and must present herself as such, so Marissa kept the crown on her head.

Merlin was once again at her feet, and she welcomed his company. The events of the past two days seemed surreal, and his presence helped to make her feel grounded, as she wandered among the revelers.

Sir Erick was always within sight, but he kept a respectful distance. She wished he would socialize with her, but she knew he was extremely conscious of protocol.

Forsythe, however, seemed oblivious to it. He walked straight up to her, bowed quickly, and reached for her hand.

"So, my queen," he said, "congratulations on your victory. The stories are already being told of the Golden Queen who defeated the experienced old warrior. The stories tell how he fled in terror, crying, 'Retreat!' to his soldiers. Pity I could not have witnessed it myself."

"Yes, pity," she replied dryly.

He seemed not to notice her disdain, and continued. "From what I hear, it was magnificent to behold. The way you sped across the marsh on your white mare, her hooves not even touching the ground, looking like an avenging angel. And then wielding the Dragon Slayer and driving the old king to his knees. Oh, how I wish I could have been there," he gushed.

"Something tells me, Forsythe," she said, "that open battle is not in your blood."

"Ah, yes, dear Queen. Alas, but I am no warrior. My talents lie more to the philosophical. Should I have chosen the course of battle, however, I am certain I would have performed exceptionally."

"No doubt," she said, not even trying to hide the contempt in her voice.

He still refused to acknowledge her scorn, and continued talking. "You have impressed the people of this kingdom immensely. Each and every subject of Avonridge would willingly do battle to protect and honor you."

"That is a very nice sentiment," Marissa said icily.

"And completely true," he assured her. "I myself would do whatever was necessary to keep you safe."

"Well, let us pray it never comes to that," she said.

"Of course, my queen," he quickly agreed, smiling.

Marissa nodded, trying to think of a way to get away from this conversation, when Forsythe reached into his green waistcoat and withdrew a pure white rose.

Extending it to her, he said, "This is for you, Majesty."

She looked at the rose, then looked at him. For the first time since she had met him, she didn't see the forced smile. This time it actually looked genuine, and she was pleasantly surprised. She reached for the offering, and smiled back at him. Before she could say anything, however, he bowed and walked away. Marissa just stared after him, perplexed.

Since her first meeting with Forsythe, she had been suspicious. The fact that both Queen Gertrude and Ruth showed a dislike for him had seemed like reinforcement of her feelings, but thinking back now, she couldn't be sure. Had she possibly colored her own feelings about him from what she saw in the other two women? It would be prudent to take their opinions into account. After all, they would know him much better than she.

And he had stated his intentions of marrying her the very first time they had met. He obviously had high aspirations. But perhaps she didn't need to be so openly hostile toward him. She would never in a million years consider marrying him, but perhaps she could consider enjoying his company. And she had to give him credit for being honest about his intentions. Even though he had colored it somewhat, he told her he intended to become king. Of course, that dream would go unfulfilled, but his company might not be so offensive after all.

Marissa thought about Sir Erick. He was still standing off to the side, keeping a close watch on her, but that was all he ever did. She would love to get to know him better, but he knew his place and would never overstep his boundaries. As frustrating as that was for Marissa, it also comforted her. She had full and complete trust in him, and was glad that it wouldn't be compromised. But, on the other hand, Marissa could certainly use some male companionship.

The past several days had been so hectic, she hadn't really had a chance to get herself settled. Hopefully now she would be able to catch her breath before the next "challenge" arose, and it would be nice to have a conversation with someone who wasn't completely in awe of her.

She began walking through the crowd of revelers, absently twirling the rose between her fingers. People bowed or curtseyed, calling greetings, but then hurried off again. It seemed as if she was an inconsequential part of the celebration, even though the celebration itself was for both her coronation and her victory in battle.

She smiled to herself and thought, let them have their fun. For all I know they may not get a celebration other than when the new queen arrives.

That was another question she wanted an answer to; how long had Gertrude reigned? And how was she chosen if the queen usually had no children? She went in search of Ruth, who, so far, seemed to have most of the answers at least.

"That is true, Majesty," Ruth said, "the queen has no children. Childbirth is so very dangerous, we want to spare her. Of course, it is entirely up to the queen herself. If she should decide to marry and have children, well, she is queen and may make her own decisions."

"And has that ever happened?" Marissa asked. "Has a queen of Avonridge ever married and had children?"

"Not to my knowledge," Ruth said. "As you know, Queen Gertrude did not. But I was not here for her coronation, so I can only tell you what I have been told."

"Yes, of course you weren't born yet. Gertrude must have reigned for forty years or more." Marissa said off-handedly.

"Oh no, Majesty," Ruth quickly replied. "Queen Gertrude ruled for four-hundred, sixty-three years."

Marissa stared at her companion, not immediately comprehending. She must have heard incorrectly. "How many years?" she asked, certain she would hear a more reasonable response.

"Four-hundred, sixty-three years," Ruth said again, and waited for Marissa to absorb that information and ask any necessary questions.

"But…that's not possible," Marissa said. "She was only sixty-five or seventy."

"No, Majesty. Queen Gertrude was five-hundred years old when she died."

Again, Marissa stared at the young woman. How could that be? She had seen Gertrude herself. There didn't seem to be anything supernatural about her, she seemed as normal as anyone else. But then again, this whole kingdom seemed strange and supernatural. Her own cat, whom she'd had for fifteen years, seemed to be changing right before her eyes. She shook her head and tried to figure out what this meant.

Gertrude had reigned for four-hundred, sixty-three years. She was five-hundred years old when she died. Marissa was no math whiz, but she figured out Gertrude was thirty-seven when she ascended the throne; the exact age Marissa was now.

"And the queen before Gertrude?" she asked. "How old was she when she died?"

"From what I have been told," Ruth said, "she was five-hundred years old when she died."

"And the queen before her?"

"Five-hundred years old."

"How far back can you go?" Marissa asked.

"If you like," Ruth said, "You may read the histories. They will have all the information you need."

NINETEEN

The room was cool and dark, lit only by torches blazing on the stone walls. Marissa walked through the room, Merlin trotting silently at her feet. The satin skirt of her gown made whisper-soft swooshing noises on the hard stone floor. Set deep beneath the rooms of the castle, she wondered why it didn't feel damp and musty. Surely, the cold and dank of a dungeon were not good for old books, yet this room didn't feel like a dungeon at all. Except for the fact that it had no windows, the atmosphere seemed almost welcoming.

Walking behind the little gnome-like man that was the Keeper of the Histories, Marissa could feel the same sensations as when she wore the golden armor; the room seemed to be speaking to her. She could feel a sensation of being included, making her a part of the room, or rather, the spirits that were housed within the room. She felt a kinship, a sense of belonging.

The Keeper of the Histories, Stefan, led her down a walkway between rows of shelves. The shelves held hundreds of old tomes, seeming as old as the world itself. The pristine condition of the old books amazed her, there was not a speck of dust anywhere. Obviously, Stefan took his work very seriously, and it showed.

Stefan stopped at the back of the room, in front of a wooden door. He opened the door, and stepped to the side, allowing Marissa entry. She walked through the door into a tiny room that was occupied only by a small table and one chair. Upon the table blazed a large candle, giving ample light to see. Next to the candle lay a leather-bound book, eighteen inches in length, a foot wide and six inches thick.

Marissa looked at Stefan, whose wrinkled face looked as if it would crack with his smile, and he nodded his grizzled head.

"The Histories, Majesty," he croaked in a hoarse voice, then hobbled back the way they had come.

She sat down on the chair, and Merlin jumped onto the table, curling up next to the book. Marissa pulled the tome closer and looked at the cover. Engraved in the old leather, etched across the center, was the inscription, "Avonridge and Her History." She ran her fingers over the carved leather, feeling the letters. The leather felt soft and supple, not old and dried, as she had expected. The leather was dark with age, but had been oiled and cared for its entire life.

Gently, reverently, Marissa opened the cover. As she drew it back, she could feel the stirrings of spirits, kindred spirits, welcoming her home. Opening the book to the first page, Marissa began to read.

The wars have been fought for ages, no one knows when they began, no one knows when they will end. King Blackburn of Avonridge fights well, but rules poorly, as have all previous kings. His main concern is his army, leaving his subjects to fend for themselves. With every able-bodied man of Avonridge serving King Blackburn, there are very few crops grown to feed the people. What crops are grown are taken for the army. Starvation threatens the people of Avonridge, and King Blackburn does nothing to abate it.

The neighboring kingdoms fare no better, as all the kingdoms are at war with each other. It has been this way for longer than anyone can remember. Generations have passed down stories of battles won and lost; babies born and grandparents dying; summer rains and winter snows; and always the wars continue. It is the way of men to fight, and Avonridge has always been ruled by men. Until that day when Alice came.

Unmarried at age thirty-seven, Alice was approaching spinster-hood. She had no love for men and their warring ways, having witnessed the consequences of war throughout her entire life. Alice had grown up with no father; he had died in King Blackburn's army. Her brothers had followed their father, and she had grown to hate the arrogance of men and their wars.

Alice saw the devastation of war as she watched her people and her land fall into ruin. The king that should be attending to his subjects had been away for years, allowing his people to starve. She vowed to correct this grievous wrong, and sent a message to King Blackburn to put an end to the war and return to rule his people. King Blackburn responded, denouncing Alice's pleas as treason. He ordered her to be brought to him at the battlefield, where he announced he would behead the woman for speaking against her king.

Alice obediently went to the battlefield to meet her king face to face, determined to plead her case for the people of Avonridge. At midnight she drew near King Blackburn's camp. When she arrived, however, a great fire blazed through the sky. The fire was a dragon's flame, as a giant beast circled above King Blackburn's army. The soldiers ran to their king in terror.

As Alice approached King Blackburn, the dragon landed on the ground. King Blackburn turned to face the dragon. The soldiers all drew back away from the dragon, as their king prepared to do battle with the beast. King Blackburn raised his sword and charged the dragon. The dragon whipped out his tail and, in one motion, decapitated King Blackburn.

As the soldiers witnessed the destruction of their king, they fled in all directions. Alice stood watching the soldiers flee, then looked at her fallen king. In his hand was still clutched his sword, and at Alice's feet lay a discarded shield. She quickly grabbed the shield as the dragon blazed fire at her. The flame raced across the ground, consuming the body of King Blackburn, but Alice stood safe behind the shield. When the flame ceased, all that was left of King Blackburn was his golden sword. Alice quickly grasped the sword, prepared for the heat to sear her flesh, but was surprised to find it cool to the touch.

She brandished the sword and rushed at the dragon. Again, his tail whipped out, but Alice dodged it and bore down on the awesome beast. As she drew near, the dragon slashed with his claw, but Alice brought the sword up and sliced off the dragon's finger. The dragon roared in pain, then leapt into the air and flew back from whence it had come.

The soldiers saw the dragon's retreat, and rushed to cheer for Alice. With King Blackburn dead, Alice now saw a way to save her people. She raised the sword in the air, declared it the Dragon Slayer, and by virtue of having defeated the dragon, claimed the crown. She brought the soldiers back to Avonridge and engaged them in farming. Within two summers, the crops once again flourished and the people had enough to eat.

The wars that seemed to have gone on forever did not encroach upon the borders of the now-peaceful Avonridge. Alice did keep an army for the protection of the kingdom, but years passed before its services were needed again.

Marissa sat back and exhaled. So, this is how Avonridge came to be ruled by women. As always, men were only concerned with playing army. It took a woman to ensure the survival of her people and keep them safe.

And Alice had been thirty-seven years old. Marissa could see the pattern here. She wondered how long ago Alice had ruled. There were no dates in the Histories, so she couldn't be certain. But at least the question of how the Dragon Slayer had come to be had now been answered.

She scratched Merlin behind the ear, then turned the next page and continued reading.

During the reign of Queen Alice, many changes came to Avonridge. The old castle was in disrepair from so many years of the king's absence. Queen Alice tore down the old stones and erected a new palace in their place. The moat had grown stagnant and foul, and Queen Alice had it drained and refilled. The parapet walls were rebuilt and fortified. When Queen Alice was finished, Avonridge looked new again.

Palace staff was gathered, ladies-in-waiting were chosen, and pages were brought into knightly service. The kings that had waged war with all neighboring kingdoms for so many generations seemed to completely forget about Avonridge. Until, that is, the day King Ragmere declared war on Queen Alice.

Having seen the prosperity of Avonridge, King Ragmere grew jealous and wanted to claim it for his own. While he gathered his army and began to march on Avonridge, Queen Alice called upon the goldsmith to fashion a suit of armor. With the armor created, she donned it and went in search of the old seer, Camille.

Camille infused the golden armor with the spirit of Avonridge, and Queen Alice rode out with her army to face King Ragmere. With the strength of her kingdom's spirit alive within her armor, Queen Alice quickly defeated King Ragmere, thereby ensuring the protection of Avonridge.

Marissa rubbed her eyes. Enrapt by the story of the first queen of Avonridge, she hadn't noticed the passing of time. With no windows in the room, she had no idea what time it was, but her aching muscles told her she'd been sitting for hours.

She thought about the golden armor. She could feel the spirit of it each time she wore it. It spoke to her, giving her support and guidance. Camille had infused the armor with that spirit, but who was Camille? The History had called her "the old seer," but didn't tell anything more about her. She would have to remember to ask Ruth.

Too engrossed in the History to consider leaving it now, she continued reading.

Queen Alice's victory over King Ragmere fully instated her as Sovereign of Avonridge. She ruled with love for her people, always seeing to their needs. As the years passed, the beauty and prosperity of Avonridge grew. The respect and love for Queen Alice also grew, and most neighboring kingdoms were in awe of her.

Several years after her defeat of King Ragmere, however, another king wished to steal the throne of Avonridge. Again, Queen Alice dispatched the attacking army and emerged victorious. Word of her wartime prowess spread throughout the lands, and no other king dared challenge her. Queen Alice lived the remaining years of her reign in peace.

As time passed it became clear Queen Alice was touched by magic. As those around her aged at a normal pace, Queen Alice seemed to remain forever young. It appeared she may rule for eternity. Her subjects were happy to have her remain, as they could remember the poverty and famine of when men ruled. Those that were too young to remember were told the stories of constant warfare and starvation. Knowing Queen Alice had had the strength to lift them up from that lowly state, they wished she would rule forever.

After four hundred years of ruling Avonridge, however, Queen Alice began to fade. Gray hairs appeared in her soft, brown locks, and wrinkles grew around her sharp eyes. It seemed her reign was coming to its end. Now arose the question of who would succeed her.

Queen Alice had never married, and so had no legal heir. The thought of countless combatants fighting for her throne, and her beloved people being ruled by a war-minded man made her heart ache, and she knew she must prevent that at all costs. She called her trusted advisors to her chambers, and together they devised a plan to insure the instatement of a worthy successor.

Queen Alice sent scouts out to the people of Avonridge. They would do nothing but watch, and report back to her. Soon, in Queen Alice's four hundred sixty-third year, the news she had been waiting for arrived. A young woman was about to give birth. This young woman was newly married to a man with no siblings, and she herself was an only child. The babe would have no blood relations other than its parents.

Queen Alice had the woman and her husband brought to the palace. The child, a girl, arrived presently with no complications. Queen Alice bestowed upon the child the name Celeste. Celeste and her parents remained at Castle Avonridge for one year, then they returned to their village. During that year, the old seer Camille blessed the babe daily, washing her head in scented oils and speaking words into her ear.

In her home village, Celeste grew to womanhood, totally unaware of her destiny. Her parents were charged with keeping silent about Celeste's first year, and they obeyed willingly. When Celeste turned eighteen, she married. After several years, during which she bore no children, her husband was struck down by a fever and died, leaving Celeste alone at age thirty-seven. She despaired, feeling desolate.

Shortly after being widowed, a knight, Sir Parker, arrived at Celeste's door. He carried the suit of golden armor and the Dragon Slayer. Celeste tried to dismiss him, but Sir Parker insisted. She acquiesced and donned the golden armor. Sir Parker taught Celeste how to wield the Dragon Slayer effectively, and together they rode for Castle Avonridge.

Upon Celeste's arrival, Queen Alice and all of Avonridge rejoiced. The successor to the throne had arrived, and Queen Alice knew her people would be well cared for. The day after Celeste arrived, Queen Alice breathed her last breath. She was five hundred years old.

Marissa gave an involuntary shudder. *So, this is how the successive queens of Avonridge come to power.* Marissa's story seemed to duplicate Celeste's except for minor details; they were both alone with no children, both were themselves only children with no other blood relatives, neither knew of their destiny before the knight found them and taught them. It was an eerie feeling, seeing in the Histories a story so similar to her own.

Marissa wondered how Alice knew Celeste would rule well. If the only criteria for a successor was a woman alone, how could she be sure the child would grow up to bear no children? And when Alice brought Celeste's parents to the palace, she only had a fifty-fifty chance of the baby being a girl. Alice had placed all her hopes, and all the hopes for the future of Avonridge, on a baby. What if she had been wrong?

Merlin lifted his head and stared at Marissa. She looked into his eyes that seemed so human, and asked, "How did she know?" When he didn't respond, she wasn't sure if she was relieved or disappointed. She'd almost expected him to speak to her.

Wanting to continue reading, but being too stiff and sore from sitting for hours, she begrudgingly closed the book. She stood up and stretched, trying to relieve the ache in her shoulders. Merlin did the same, purring loudly as he arched his back. He jumped to the floor and followed Marissa as she walked out the door.

TWENTY

Marissa stepped out into the courtyard with Merlin tight at her heels. The delicious smells of roasted meats filled the cool evening air. The celebration continued, and people milled about drinking and laughing and eating. As always, Sir Erick waited for her to emerge, and he stood nearby. Marissa, however, wanted to find Ruth, as she seemed to have most of the answers.

She waved a greeting to Sir Erick, who nodded in reply, then went in search of Ruth. Her stomach growled, reminding her she hadn't eaten in several hours, but that would have to wait. Right now, she needed exercise and information.

Ruth wasn't difficult to find, as she never strayed too far from Marissa either. She found her standing off to the side, chatting and laughing with a group of young women. When Ruth spied Marissa, she quickly broke from the group and hurried to her queen. The young women also rushed to Marissa, dropping quick curtseys and calling, "Good evening, Majesty."

Marissa greeted the young women, then said, "Please excuse us ladies, I need to speak with Ruth." She grasped Ruth's arm and together they walked away, leaving the group of young women very obviously disappointed.

When they were out of earshot of everyone around, Marissa said, "Ruth, I've been reading the Histories for several hours, and they're simply fascinating."

"Yes, Majesty," Ruth said. "The story of how Avonridge came to be is indeed captivating. I knew you would find what you were looking for there."

"Well, I found the answers to some questions there," Marissa agreed. "However, I also found even more questions. For instance, how did Queen Alice know Celeste would be a good ruler? Did she already know Celeste's parents? From what the Histories said, it doesn't appear so."

"I am sorry, Majesty," Ruth said, "but my knowledge is quite limited. Only the queen is permitted to read the Histories, so my information consists of the stories passed down through the generations."

"Well, is there anyone who might know more of the way things happened? The Histories tell what happened, but not a lot of whys."

"Perhaps you would like to speak to the old seer, Camille," Ruth offered.

"Camille?" Marissa echoed. "But she can't still be alive after all this time?"

"I am sure whatever questions you have, she can answer them."

Marissa followed Ruth along the wooded path. Merlin trotted at her heels and Sir Erick marched two steps behind. The forest through which they traveled was lush and fragrant. The sun was just rising and it made the dew-covered foliage glisten and sparkle like diamonds.

Long strands of rich moss clung to the tree branches, hanging down in emerald green cascades. Lichen grew on trunks giving the trees a velvety look. Great green ferns were scattered here and there, looking like huge leafy flowers in bloom. The path was littered with fallen leaves, as the trees were beginning to prepare for the winter.

The little group moved along silently, each enjoying the richness of the morning. The air was still cool, and Marissa breathed deeply, filling her soul with the life and spirit of this mystical place.

Ruth continued to lead the way, and eventually the little group emerged into a small clearing. Off to the side of the clearing sat a little hut made of small logs and sticks laced together with strands of ivy. The hut stood approximately six feet high and eight feet across. The top was rounded and the base seemed to be a perfect circle. A doorway was fashioned into the front and a large animal skin hung across it.

To the left of the hut, back against the tree line, sat a stone bench about four feet in length. To the right of the hut was a small firepit with a huge iron cauldron braced above the fire. A small wooden frame stood next to the fire, holding two huge wooden spoons and one large ladle with an extremely long handle. Steam was already rising from the cauldron, but no one was tending it. Marissa's imagination conjured up all sorts of images of eye of newt and toe of frog simmering away in the huge pot, but she pushed those thoughts aside. No need to prejudge anyone; she needed to speak with Camille in a serious tone.

The little group waited just inside the clearing. Marissa looked at Ruth, prepared to ask a question, but Ruth silenced her with a quick gesture. Marissa begrudgingly obeyed, and waited quietly. After several moments, the animal skin over the hut doorway was pushed aside. It was too dark to see the interior of the hut, so Marissa and her party waited a few moments. After several heartbeats, Marissa decided that the inhabitant of the wooden hut was waiting for her, so she spoke.

"Camille," she called. "It is I, Queen Marissa. I wish to speak with you. Will you grant me the honor of your presence?" Knowing this woman had the awe of the people of Avonridge, Marissa decided to show respect instead of making a demand. As the seer, Camille was believed to have some magical powers. Whether she possessed magic in actuality would remain to be seen.

From within the hut Marissa could finally see movement. An old woman emerged from the doorway and stood up. She was tall and thin, and the moment Marissa spied her, her determination not to prejudge evaporated. Long, white hair hung down the old woman's back, thin and wispy. Although it was unkempt and unruly, it was obviously clean. As were the rags she wore. Torn and tattered and of myriad colors, strips of cloth were tied around her waist, crisscrossed over her shoulders, hanging like veils from other strips of cloth, making her look like a battered walking rainbow.

Despite being old and thin, the woman stood straight as a ramrod. Her shoulders were gaunt, but pushed back in a self-assured manner. Her hawk-like face was withered and wrinkled, but her eyes were keen and sharp, even across the distance of the clearing. The woman looked directly at Marissa, her countenance showing respect but not subservience.

"Are you Camille?" Marissa asked.

"I am," the old woman replied in a clear, strong voice.

"I wish to speak with you," Marissa said.

Camille nodded her grizzled head. "Of course, Majesty," she said. "I am at your service. Would you sit?" she asked, indicating the stone bench at the side of the clearing.

"Thank you," Marissa said, as she strode to the bench. Ruth, Sir Erick and Merlin all followed their queen.

Upon spying the little black cat, Camille's face broke into a broad smile. "Why Merlin, it is good to see you, my friend."

Merlin stopped and turned toward Camille, meowing loudly in obvious reply. Marissa looked at her little companion and shook her head. Again, she wondered just how he fit into this whole situation.

When Marissa was seated and Ruth and Sir Erick were standing to either side of her, Camille walked to her and asked, "Majesty, may I get you some refreshment? I know you've been walking for a while."

"Thank you, but no," Marissa answered. "I really would like to get as much information as I can."

"Of course, my queen," she said, and sat beside Marissa. "What is it you wish to know?"

"Well, first of all, are you a direct descendant of the old Seer, Camille, who helped Queen Alice?"

"No, Majesty," she said with a twinkle in her sharp eyes.

"You're not?" Marissa asked, somewhat perplexed. "That seems so strange. I would have thought that, with the same name and abilities, you would be some great-granddaughter way down the line. Are you related to that Camille at all?"

The old woman smiled. "That depends on what you consider 'related'."

"I'm getting confused," Marissa said. "Of course, related means connected by blood in some way. A distant cousin perhaps, or niece."

"No, I am no cousin, nor niece of the old Seer Camille who served Queen Alice."

"Then I don't understand," Marissa said, becoming impatient. "If you are not related to the Camille of old, then how can you have her powers? The name is of little consequence, but I do know the 'sight' is a rare gift. It usually stays in families."

Camille continued to smile, her gray eyes twinkling, but she took her time in responding. Ruth now grew impatient, and stamped her foot.

"Camille, Her Majesty does not have time for your foolish games," she admonished. "We've walked a long way and we are tired. Do not further burden Her Majesty with your antics. Simply answer her questions."

Camille chuckled softly, but nodded and said, "Of course, dear Ruth, of course. No need to ruffle your feathers."

Ruth sucked in her breath, but said nothing. Camille then turned her attention back to Marissa, and said, "My queen, I am not a descendant, direct or otherwise, of the Camille of old. I am the Camille of old."

Marissa looked at the smiling old woman. "You are the Camille of old," she repeated. "But how can that be? That would make you..." She faltered, not even venturing a guess at the age of the woman who sat next her.

"I am old, indeed," Camille agreed with a smile. "How old exactly, I cannot say. I witnessed the birth and death of several kings before Queen Alice rescued Avonridge."

"But how?" Marissa asked. "How could you have lived so long?"

"You have seen many things upon returning to Avonridge that have amazed and confused you," Camille said. "You are stubborn, indeed. That can be good, but too much stubbornness can prove harmful. Let go of your fixed notions of what is impossible, and view your kingdom with less question. Then you will be able to understand more, and remember more."

Marissa listened to her words. She nodded and closed her eyes, opening her mind to what the seer had said. After a moment, she said, "You've seen many kings of Avonridge? Did they live extraordinarily long lives, as did the queens?"

"The men were simply men," Camille replied, "living the expected lifespan of the era. But Queen Alice, ah, she was different."

"Yes," Marissa agreed. "The Histories said she was touched by magic. Tell me about that."

"The Histories were written by subjects who witnessed the events, yet they could not know the innermost details. They could not know how the events came to pass, why they occurred."

"That's exactly what I want to know," Marissa said. "I want to know how and why these things happened."

Camille nodded, then continued. "Yes, Queen Alice was magical. But not touched by it, she was fashioned from it."

"Fashioned from it?" Marissa repeated. "What do you mean?"

Camille gazed off into the forest, a faraway look on her face. She seemed to be seeing another place and time. Marissa waited patiently.

"When Alice was born," the old Seer explained, in a voice that was eerie and filled with emotion, "the kingdom had been at war for generations. She never knew her father, and barely knew her brothers.

They would all die in service to King Blackburn. Her mother was a wise woman, and knew the kingdom was near collapse.

"I too had witnessed the destruction warfare brings. I had grown tired of it, tired of seeing the people of Avonridge wither and die like leaves in the autumn winds. Good people were starving to death, while battles raged on. I am sure Blackburn no longer knew exactly what he fought for, having spent so long in mindless battles, the reasons were long forgotten.

"But Alice's mother, Katrinka, knew the reasons did not matter. A man who starves to death for a just cause is equally as dead as a man who dies for folly. The wars must be stopped and the people of Avonridge must be saved. She came to me one winter morning, wearing not much more than the rags I now wear. She carried a snow hare she had trapped, and presented it to me in payment for my services, any service I may see fit to offer. She was a pitiful creature, skin and bones, in the twilight of her life, and I insisted she share the hare with me."

"That evening, as she slept, I called Katrinka's husband back from the battle. He lay with Katrinka in that very hut," she pointed to her wooden home, "while I danced and sang in the moonlight. When the sun rose, he returned to the battlefield, and was cut down that very day."

Camille hung her head in painful remembrance, her white hair draped around her sad face.

"Katrinka's heart was broken," she whispered. "But she returned to her grown sons, who were preparing to go off to war to avenge their father. She entreated them to remain with her until her mourning time had passed, and they agreed.

"Presently, however, Katrinka grew great with child, the child that was created in that very hut. That child was Alice. When Katrinka's labor commenced, she sent her sons for me, and I brought Alice into this world. I stayed with Katrinka, Alice and Katrinka's sons for one year. During that time, I blessed Alice daily, bathing her head with sweet oils and whispering to her all that she would need to know to save her people.

"The rest you know from the Histories," Camille said, finally bringing her gaze back to Marissa. They sat in silence for a while as Marissa absorbed what Camille had told her. Alice was not touched by magic, she was fashioned from it.

"You said Katrinka was in the twilight of her life," Marissa said. "And her sons were grown."

"Yes, she was well past the years of bearing children. And her sons were grown men, on the verge of seeking wives."

"But they never married?"

"No," Camille said. "Instead, they sought the adventures of war."

"So, Alice was raised alone by an elderly mother," Marissa said, more to herself than to Camille. "And Katrinka shouldn't have been able to bear another child at her age. So, Alice really was fashioned from magic."

"As were all the successive queens of Avonridge."

Marissa looked at Camille. Seconds passed before she could form the next question: Did that mean that Marissa herself was fashioned from magic? Camille smiled and nodded, her gray eyes shining in the early morning light.

"That is true, my queen," she said softly, without the question having been spoken. "*All* the queens of Avonridge have been fashioned from magic, including you."

"But that can't be," Marissa argued weakly. "I don't feel like I'm magic. I've led a pretty mundane life. Until a couple weeks ago, that is."

Camille nodded again. "As it is with all queens of Avonridge. They know nothing of their destiny until it is their time."

"But the Histories mentioned nothing of the next queen, Celeste, was it?" she said, trying to remember exactly what she'd read. "It said nothing of Celeste's parents coming to you to conceive the child. All it said was that the child was already on the way when they were brought to court."

"As I said, the Histories were written by witnesses of the time. They wrote what they knew, and so there are many gaps."

Marissa nodded, accepting that as quite possible. "But why then, are the parents and babe sent back to their villages? Why aren't the new queens raised at court? That way they would grow up knowing what was expected of them. They could be instructed in the proper way to behave, and be taught how to fight at an earlier age. I just learned how to wield the Dragon Slayer only after Sir Erick found me. Wouldn't it have been better to teach me earlier?"

"Did you feel unprepared when you met King Sorens?" Camille asked.

"Well, no," Marissa admitted.

"Then to what advantage would it have been to teach you earlier?"

Marissa thought about that for a few moments, then acquiesced and nodded.

"A young girl growing up at court," Camille said, "knowing she would one day become queen, that information can be quite a heavy burden. She may be anxious for it to occur, or she may even be apprehensive at the thought of the responsibility involved. There is also the very real possibility that, were she known to the kingdom before her time to ascend the throne, her life could very well be in danger.

"And you brought a fresh viewpoint to the battle," she continued. "Had you been more prepared for it, you may have grown complacent. As it stands, you were thrown into your challenge without the burden of prior opinions, and you performed admirably. Your magical origin and my blessings and words were all the preparation you needed. So, tell me, my queen, what would you have changed?"

"I suppose you're right," Marissa agreed. "It's still just so strange. I was conceived and born here in Avonridge, then sent to live in...," here she faltered. She had originally thought of her home as the 'real world.' But as the days passed, she came to feel like Avonridge was truly her home, and the term 'real world' seemed so subjective. Avonridge certainly felt just as real as her old home. In fact, looking back now, she could easily convince herself that her previous life was nothing more than fantasy.

"Your parents were older," Camille said. "And you were an only child."

Marissa nodded. She had grown up loved, but had no family to speak of.

"Did your parents ever tell you of your ancestry?" the old woman asked.

"No, they never told me anything. When I would ask, they'd simply tell me we had no relatives. I thought it strange as I got older, most of my friends had aunts and uncles and cousins. But that was just the way our family was."

"Yes, you were chosen even before you were conceived," Camille stated.

"I don't understand," Marissa said.

"Perhaps I should say your parents were chosen. Just as it says in the Histories, knights were sent out in search of the couple that would bear the new queen. Your parents were chosen for their intelligence, honesty, strength, and lack of familial ties. Much was the criteria that needed to be met. When your parents were found, they were entreated upon to come to Avonridge. Being good people, wise and trusting, they agreed. They were brought to me, and they spent the night in that hut," again she indicated her little home.

"As with all previous conceptions, I danced and sang in the moonlight. And as with all previous conceptions, your spirit came down from the heavens and became the child that would be you."

Marissa sat quietly, overwhelmed by the information Camille had imparted. Her spirit came down from the heavens. Well, having been raised Christian, she already believed that much, but Camille was telling her she was magical. Yes, her parents were older when she was born. They'd told her they didn't believe they would ever have children, and were overjoyed when she came along. But women had babies later in life all the time.

And how could a knight convince a couple from the outside to come to Avonridge in the first place? If she wanted to hold onto what sanity she still possessed, Marissa should do exactly as Camille had instructed; she had to let go of her old-world ideas and look at Avonridge more openly. If she tried, she could question and doubt everything, but to what end? She already knew this place held mysteries; she'd fought a dragon after all. And she had been crowned queen. What more could she ask for?

Camille allowed Marissa to reason things out for herself, and sat beside the queen in silence. Her grizzled old face had a look of serenity and patience that Marissa found comforting. Finally, Marissa nodded and stood up.

"Camille," she said, "I thank you for all you've told me. It is interesting, and helps to put things into place. I will take your advice and look at my kingdom with an open mind and an open heart." She turned to go, and Ruth and Sir Erick fell into step with her. Merlin jumped into Camille's lap, and she hugged the little creature to her breast.

"Yes, my friend," she said to the cat. "You must visit more often. It is good to be graced with your company." He meowed and jumped down, trotting to catch up to the little group.

They had almost reached the opposite side of the clearing when Camille called. "Just one more thing, Majesty."

They all stopped and turned to face the old seer.

"Beware. Treachery lives."

"Treachery," Marissa repeated. "But who?" But Camille didn't answer. Instead, she returned to her hut and pulled the animal pelt back across the doorway. This visit had obviously ended.

"Well, that's annoying," Marissa said. "After all the things she's just told me, why wouldn't she tell me who the traitor is?"

"Camille is quite old," Ruth said. "Perhaps her mind is beginning to show the failings of her age."

Marissa looked at her companion in amusement. Ruth certainly seemed awed by the old seer, and perhaps a little leery of her. Marissa shook her head and turned toward home.

TWENTY-ONE

The celebration was finally winding down and the people were returning to their villages. Hopefully now, life would return to normal at the palace and Marissa would at last find out exactly what normal life at the palace was like.

She began wandering through the castle, learning its layout, discovering different corridors and rooms. She began with her own wing, and was surprised to find other rooms beside her own chambers. She hadn't noticed the doors on her several trips up and down the corridor, but she'd been a little preoccupied on those trips.

There was a sitting room with fresh flowers placed around in large urns, which had the sweetest scent Marissa had experienced in a great while. Off to one side was another room with a table that looked like it was used for playing cards or board games. She continued walking down another corridor and found room after room that seemed to be duplicates; all large and furnished much the same except for the color scheme.

There was a royal blue room that had a blue velvet settee and blue velvet chairs scattered about. The walls were draped in blue, and blue flowers sat in glass vases upon small tables. An emerald green room was next to the blue room, with the same furnishings and draping, with potted ferns instead of flowers. Beside the green room was a red room, with blood-red roses in vases on tables.

Marissa turned yet another corner and found a great drawing room, warmly furnished with gray velvet settees, and a large harpsichord.

Having always wanted to play the piano, she decided to have someone teach her the harpsichord.

As she walked, she saw the banquet hall in which she had first met Queen Gertrude. It seemed to be the all-purpose meeting room. The throne sat at the back of the hall on a small dais, and the colorful draping still adorned the walls. Although there were no people in the hall at the moment, it still appeared well tended. The stone floor gleamed, and the walls were freshly cleaned.

She wandered slowly through the hall, remembering her first meeting with the dead queen. Marissa had been welcomed by the whole kingdom that day. Well, the whole kingdom minus one inhabitant. Marcellus was still a thorn to be dealt with. His presence could be felt everywhere, making her feel a constant need to look over her shoulder. It was clear he wanted the throne, but to what lengths would he go?

Sir Erick had said it was believed Marcellus had summoned Balvindor. Marissa did not believe the old wizard and the dragon were allies. In fact, she believed Balvindor would not accept Marcellus as sovereign. Balvindor was a respectful creature, as dragons go, but Marcellus seemed just the opposite. Marcellus had an air of treachery about him. Treachery…

Could he be the one Camille was warning her about? It would certainly make sense. He had already told her he didn't think she should have come back to Avonridge. He wanted the throne, and she had to make sure he didn't succeed. He obviously had some magical powers, but so, it seemed, did Camille. And Marissa felt she could rely on Camille if it became necessary. She only hoped it wouldn't become necessary to fight magic with magic.

She had to keep watchful and not allow the old man to surprise her. He apparently could slip in and out of the castle unnoticed. But she could sense his presence. She could tell if he was near, and she had to be aware of that feeling if she was to stand against him.

She continued thinking about Marcellus and how uneasy he had made Sir Erick on Marissa's initial trek into Avonridge. She thought about how Merlin had arched his back at the old man, and how she challenged him even then. As she walked through the corridor, Ruth came hurrying up to her.

"Majesty," she said, and dropped a quick curtsey, "Roland wishes to have a word with you."

"Of course," Marissa replied. "Please send him to my sitting room."

"Yes, Majesty." And the young woman hurried away.

Marissa began walking back to her chambers, welcoming the idea of seeing Roland again. Since the celebration for her coronation and battle victory had ended, she was almost feeling like she was no longer needed. Palace life was very quiet. She supposed she should feel lucky, even grateful, that life was sailing along so smoothly now, but she was beginning to get bored.

Never having been what one might call a social butterfly, Marissa hadn't had a great deal of friends during her lifetime. She'd always had one or two good friends to pal around with, but she never went in for the big social scene. Constantly having a host of girlfriends over, or going to this one's house or that one's house just never caught on with her. In fact, she was quite the loner when left to her own devices.

Now, within the palace, she had ladies-in-waiting. A half dozen or so, she wasn't even really sure exactly how many, were constantly vying for her attention. Ranging in age from about twenty to forty, she had to contend with hyper young women trying desperately to please her and therefore curry favor with the new queen. There were also women her own age who seemed to have grown cynical and jaded. It was almost a burden to have to deal with all the women, she would much rather have one friend and confidant. She more or less had that with Ruth, but again, Marissa was Queen and had no actual peers.

Now Roland wanted to speak with her and she found herself almost hoping he would be bringing news of another of her challenges. As she entered her sitting room, Roland was standing at the far wall. When he spied her, he bowed deeply.

"Majesty," he said, his voice crisp and business-like.

"Roland," she greeted warmly. "It is good to see you. Please sit and tell me what's on your mind."

He waited until his queen seated herself, and then took the chair across the little table from her.

"Majesty," he said again, and Marissa could hear slight trepidation in his voice. "As you know, the kingdoms all around us have been warring with each other for many generations."

Marissa nodded, remembering the Histories and wondering if those very same wars had been raging for thousands of years.

"It seems a High King had been trying to unite the lesser kings, and had made some progress, but was killed in battle some years back."

She hadn't known that bit of information, but nodded anyway for Roland to continue.

"The Pendragon had persuaded a few of the lesser kings to call a temporary truce. But with his death, they began their wars again."

"I know a little of the stories of the wars," Marissa said. "It seems as if the warring kings forgot about Avonridge. We're quite lucky in that regard."

"Yes, Majesty," Roland agreed. "King Sorens is the first to challenge us in my lifetime. Hopefully his act of aggression will not bring attention to us and drag us back into the wars."

"But you mentioned the Pendragon," she said. "I've heard that title, and the only Pendragon I know of is Uther."

"'Twas he, Majesty," he replied. "Uther, the Pendragon. He was killed a time ago, and now his son Arthur carries the title."

"Arthur," Marissa breathed. "King Arthur? The son of Uther, the Pendragon?"

"The same, Majesty. And I am told he is an upstart, an arrogant king bent on earning for himself the fear and respect the title 'Pendragon' brings. He rides this way, and I believe we should prepare to face him."

"Of course, of course," she gushed, and stood up to pace the floor.

Obediently Roland rose as well, and continued. "We have no idea of what his intentions are. We must be prepared to face his army when he arrives."

"Yes, when he arrives," she said, awe and excitement making her breath catch in her throat as she paced back and forth. Spinning on her heel, she faced Roland and asked, "When will he arrive? When will King Arthur be at Avonridge?"

Roland looked confused. "Majesty," he said. "You seem pleased that Arthur and his army march this way. Do you not understand he may be declaring war on Avonridge?"

"War?" she said, incredulous. "Of course he won't declare war on Avonridge. He is the High King trying to unite the lesser kingdoms."

"Yes, but by what means? He will use whatever force he feels necessary. We cannot trust this man, Majesty. Many have called him a usurper, declaring he has no legitimate claim to the Pendragon throne. His mother was married to another man when Arthur was conceived, making him a bastard child. A bastard has no right to wear a crown."

"Oh, don't be ridiculous," she said, waiving away his protestations. "Now, tell me Roland, when does Arthur arrive at Avonridge?"

He shook his head, a look of confusion and doubt on his face, but answered her question. "Two days hence, Majesty."

"Splendid!" she nearly shouted, clapping her hands together. "Have the castle scrubbed from top to bottom. Make this place glisten and gleam like gold. Use the captured soldiers from Sorens' army. They've had enough of a vacation, make them earn their keep. Give them rags and scrub brushes, and get them to cleaning and polishing. I want this castle to rival anything Arthur may have at Camelot."

Roland continued shaking his head, absolutely certain his queen had lost her mind, but obeyed nonetheless. He walked out the door to begin preparations for welcoming the bastard king, muttering to himself, "Camelot? What in Creation is Camelot?"

* * *

Just as Marissa was overseeing the cleaning of the palace and gathering of banquet foods, a young boy ran to her, crying and shouting in great distress.

"Majesty! Majesty!" he called as he hurried up to her. "Please Majesty, help my father."

The boy was no more than six or seven years old, and was on the verge of hysteria, tears streaming down his dirty face. "Please Majesty," he cried. "Help my father."

Sir Erick rushed up to him and gently grasped his arm. "Stand clear, son," he commanded, as he dragged the boy back several paces from Marissa. The presence of the Black Knight quieted the boy somewhat, either from awe or downright fear, it wasn't clear.

He gulped air for a moment, attempting to stop himself from crying. But when his gaze fell back on his queen, he began anew, practically hopping up and down, crying, "Please Majesty, please."

Sir Erick looked as if he would haul the hapless child away, but Marissa stopped him.

"Sir Erick, it's all right. Let the child speak. He obviously needs me." Then to the boy she said, "Tell me child, what has you so upset?"

"Oh, Majesty!" he wailed, almost too agitated to speak clearly.

The Black Knight, still grasping his arm, gently shook the boy and said, "Calm down, lad. Her Majesty cannot help you if she cannot understand you."

The boy calmed down to simple crying, and tried again. "Majesty, the beast has taken my father."

"The beast?" Marissa echoed. "You mean Balvindor?"

"Yes, Majesty," he said between sniffles. "He swooped down on my father while he was herding the cattle and took him off somewhere."

Marissa shook her head. She didn't want to believe the dragon had begun terrorizing the people again, not after he'd been so civilized at her coronation and the banquet. Well, she supposed a dragon couldn't change his scales.

"Now, you're certain it was Balvindor, and not another dragon?" she asked.

"Oh yes, Majesty," he said earnestly. "It was Balvindor. He called to me as he flew away."

"He did?" she asked, amazed, and the boy nodded. "What did he say?"

"He said," and the boy stood up straight, ready to recite from memory, "'Tell your new queen her challenges have not been completed.' Then he flew away with my father." He wiped his eyes with one dirty hand, then drew the back of it across his nose, sniffling loudly.

Marissa looked at the Black Knight, her heart aching for the father of this little child. Then she asked the boy, "Is your father still alive?"

"Yes Majesty," he answered. "Balvindor only carried him away."

"At least for now," she muttered quietly. Then to Sir Erick she said, "Gather a band of twelve soldiers. We will ride out in one hour."

"Yes Majesty," he said, and hurried off to get it done.

Marissa then squatted down to be on eye level with the boy. "Tell me, child," she said, "what is your name?"

"Darvin," he said, still sniffling.

She reached out and grasped his arms, drawing him closer to her. "Darvin, I want you to know I will do everything in my power to bring your father back safe and sound."

He looked at her with grateful eyes and managed a tentative smile. "Thank you, Majesty," he whispered, and then spun around and ran away.

Exactly one hour after meeting with Darvin, Marissa sat astride her white mare, Selene, clothed in her golden armor, with the Dragon Slayer strapped to her side. Grasped in her left hand was the shield that had protected her well from Balvindor's fire. Beside her sat the Black Knight atop his black steed, and behind were twelve mounted soldiers.

Roland was standing next to the queen, arguing for her safety. "Really Majesty, this happens upon occasion. You cannot do battle with the dragon over each and every man he terrorizes."

"But if I am to protect my people," she countered, "then I must protect each and every one of them."

"Yes Majesty, but you still need to…"

Marissa cut him off. "Roland, I know your only concern is for my well-being, and I am grateful for that. But I have a duty to my people, all of them. I cannot let Balvindor take my people away unchallenged."

"But Majesty," he protested weakly.

She pretended not to hear, and continued. "Now, King Arthur will be arriving in two days. Ruth will oversee the preparations, and I want you to give her your full cooperation."

He nodded, realizing he was defeated, and said, "Yes Majesty,"

"Good," she stated. "We will return as quickly as possible." And she spurred her mare and led the small procession out the castle gate and across the drawbridge.

TWENTY-TWO

"Balvindor!" Marissa called, "I have come for my royal subject."

The queen and her small band of soldiers sat atop their horses, approximately one hundred feet from the mouth of a great cave. High up in the hills of Avonridge sat the lair of the awesome beast. A mesa spread out before the cave, offering a large expanse that made Marissa think of a playing field. She waited for a reply from the dragon, but silence was all that drifted back at her.

"Balvindor!" she called again, growing impatient. "I have no time for your nonsense! Do not force me to invade your home."

"You have already invaded my home," rumbled the terse reply from within the cave. "You stand at my threshold with soldiers ready to do battle. That is indeed an invasion."

"I simply come to rescue my subject," she countered. "You have encroached upon my land and my people, I come to your threshold to set things right. Release your hostage and we will gladly leave you in peace."

"Hostage?" the dragon repeated. "Dear lady, this man is my dinner."

A chill ran up Marissa's spine as she heard a human voice cry pitifully, "No, please, no." But she was heartened to know he was still alive.

"Balvindor! I will not stand by while you make a meal of my people. There are plenty of animals for you to satisfy your hunger, you don't need to harm my people."

A deep, guttural laugh emerged from the cave causing the hair at the back of Marissa's neck to stand out. "But human flesh is so much tastier," the beast taunted. "Why should I settle for less?"

Marissa grew angry. "Balvindor!" she shouted. "I tire of this game. Your actions were obviously meant to bring me here. Well, I am here. Do you want to hide in your cave, or shall we get on with this?"

"As you wish," came the reply, and instantly he emerged from his cave. The huge, spiny head was the first to appear, and Marissa heard several of her soldiers gasp. One swore quietly. Even the horses were not immune to the awesome creature's presence, and some stamped a foot or snorted anxiously. Marissa's own mare Selene threw her head up and down several times, eager to engage the enemy.

As Balvindor hauled his immense bulk out through the mouth of the cave and up to a standing position, he towered over the members of the little group.

"Queen Marissa," he said, "you are correct in your assessment of my actions. I did indeed wish to bring you here. You have undertaken a great responsibility when you assumed the throne. But all has been going smoothly at Castle Avonridge. I do not wish for you to become complacent."

"How thoughtful of you," she replied. "However, I have much to occupy my mind. There is no need for you to pull childish pranks to keep me busy."

"Do not misunderstand me, Majesty. I fully intend to have my dinner. I simply wanted to assist you in your duties as I went about my daily activities." Then, with absolutely no warning, the dragon whipped out his tail, aiming directly at Selene. The mare, however, seemed to sense the attack, and leapt straight into the air, sailing right over the snakelike tail as if it was nothing more than an annoyance, getting Marissa out of the line of immediate danger.

The members of the rescue party, unfortunately, were not as quick to respond. Sir Erick and three of the soldiers were thrown from their mounts as the horses were knocked to the ground. Sir Erick was on his feet instantly with his sword in his hand, and Marissa unsheathed the Dragon Slayer. The horses all leapt to their feet and began to run in frenzied circles. The remaining mounted soldiers, upon seeing their comrades were relatively unhurt, hurried to intercept the spooked horses and hold them for their riders.

Marissa considered jumping to the ground, but felt a spiritual communion with her mare. The horse was trained and ready to battle the great beast, so she stayed mounted. Sir Erick was still on his feet, sword held before him in both hands, knees bent slightly, the seasoned warrior ready to protect his charge. The felled soldiers had remounted, and all fanned out in an arc around their queen. One of the soldiers brought the black stallion to the Black Knight, and he quickly remounted.

"Yes," Balvindor said appreciatively, "you are a worthy opponent indeed, Majesty."

"Enough talk Balvindor," Marissa bellowed. Her mare was stamping and snorting, eager to begin the task for which they had all traveled into these hills, and so the queen gave the command.

As her mount reared up, hooves slashing the cool air of late afternoon, she yelled, "Forward Avonridge!" and she raced toward the beast, the Dragon Slayer held high, her twelve mounted soldiers following. Sir Erick drifted back and to the side, attempting to fade in behind the dragon. Balvindor made a sweep with his right arm, trying to knock Marissa off Selene, but the mare stopped short and feinted to the side, easily evading the taloned claw. Marissa marveled at the fact that she knew exactly what the horse was planning to do. Just as she felt her golden armor communicating with her, so too she felt the mare's thoughts.

Swinging his left arm now, the dragon managed to connect with one of the soldiers and knocked him to the ground. He rolled quickly and scrambled to safety, just as a huge reptilian foot came crashing down, missing him by inches.

Soldiers with swords drawn rushed at the dragon, slashing at whatever part of the beast they could reach. Dull thuds and sharp metallic clanks began to fill the air as sword after sword made contact with the huge armored scales that protected Balvindor's body. As he twisted this way and that, the dragon made continual sweeps with his huge arms, trying to knock other soldiers to the ground, but they quickly learned to read his movements, and were leaping out of the way before he could do any damage.

The soldier that had been knocked down caught his horse and was just about to remount when he caught Marissa's eye. She made a quick movement with her head in the direction of the dragon's cave, and he nodded in response. Quickly he dashed toward the cave, and Marissa and the other soldiers redoubled their efforts to keep Balvindor's attention.

Each soldier rushed in to slash at Balvindor's leg, or foot, or lower belly, then dashed away before the beast could swat at them. Marissa herself was wielding the Dragon Slayer at the huge scales, and hearing Selene's warnings of a claw reaching for her, or the tail lashing at her. Each attack made by the little band of soldiers brought Balvindor further away from his lair as he tried desperately to catch, or at least swat away, his tormentors.

Marissa charged in and slashed at the dragon's belly, bringing her sword up in an arc that hit home. As the Dragon Slayer made contact, it was at such an angle that it slid under one of the iron-hard scales, slicing it completely free of Balvindor's body. The beast almost seemed not to notice, but Marissa paused and studied the wound. As Balvindor continued to fight off the soldiers, she could see his flesh move where the scale had been. It did not bleed, but it was obviously unprotected, and so was vulnerable. She was just about to thrust her sword into the unprotected spot when Selene cried out a warning, and leapt to the side, narrowly missing being fried by a dragon flame.

Balvindor was now using his fire, and so Marissa knew she had to bring this melee to a quick end. She dropped back and was about to regather her band of soldiers, when Balvindor bellowed out an earsplitting roar of pain and anger. The beast spun around, turning his back to Marissa and her soldiers, and she could now see what had made him roar so.

To the left of, and just beneath, the dragon's tail, buried to the hilt in the soft flesh just behind his leg, was Sir Erick's sword. The Black Knight had taken advantage of the distraction the soldiers offered, and thrust his sword into one of the very few vulnerable spots on the dragon's massive body. Unfortunately, now Sir Erick stood alone against the beast, with no weapon. All the soldiers were now behind Balvindor as the dragon sought to avenge this most distasteful of offenses against his honor.

Still mounted, Sir Erick dashed to the side to avoid being clawed, but he was between the dragon and the stone wall of the cave. He had nowhere to go, and so turned to face Balvindor and whatever punishment the dragon might inflict.

Marissa saw the grave danger her dear friend was in, and rushed forward, the Dragon Slayer held high. Her soldiers joined their queen, and, taking their cue from Sir Erick's ingenious attack, leapt to the ground to get close enough to allow their swords to join the Black Knight's. Two men ran up together, aiming at the spot under the dragon's tail that could be penetrated.

Just as Balvindor reached out a scaly claw to snatch the Black Knight from his mount, the two swords were thrust home, causing the dragon to jerk back, again roaring in pain. He spun around to face this new attack, but his left leg was now wounded and didn't move well. His left foot caught on a rock that stood about six feet high, and the great beast began to fall.

"Stand clear!" Marissa shouted to her soldiers. Selene easily carried her to safety, and she was relieved to see all of her men dash to the side, away from the tumbling behemoth. As Balvindor came down, it was like watching a stone tower fall to the ground. First his huge belly hit the floor of the playing field. Next, his chest rolled forward, and finally his chin crashed to the stone, causing a stream of fire to shoot fifty feet out of his nostrils, scorching the ground to black charcoal.

Just at that moment, the soldier emerged from the cave, half dragging, half carrying a badly shaken man in peasant clothing. His burlap trousers and rough cotton shirt were in tatters, and he had a dried bloodstain on the side of his face. The soldier helped him up onto the back of his horse, then climbed into the saddle. The man held tight to the soldier's waist, and Marissa motioned for them to ride to safety.

Balvindor lay dazed on the ground. For a few moments, Marissa wondered if he was still alive, but then his eyelids fluttered. She moved Selene to stand in front of the dragon, and called his name.

"Balvindor," she said, and waited. He groaned, and a puff of smoke billowed out of his nostrils.

"Balvindor," she said again. He opened his eyes, and glared at her. "I'm glad to see you still live," she said, with only a hint of sarcasm.

"Your concern is touching, dear lady," he muttered through teeth that had been slammed shut on impact.

"It is late," she said, "and I have duties to attend to. This is over for now. Rest up, tend to your wounds, and if you feel the need to continue this nonsense, I will meet you another day."

She then turned Selene toward the trail through the trees, and led her victorious band home.

Marissa led the little rescue party through the gate into the castle courtyard. After her previous return trips where she was met with cheers and praise, it was somewhat of a letdown to be almost ignored. People were going about their daily activities, only pausing to acknowledge their queen with a bow or curtsey, then hurrying along on their way.

What did impress her, however, was the condition of the courtyard and castle. Even in the fading light of early evening, the stones gleamed like new. The grounds had been swept clean, and men still toiled to make Castle Avonridge a sight to behold. Scaffolds were set up for men to work on washing and polishing the stones, and Marissa was glad to see Roland had indeed put Sorens' soldiers to work. Although they were kept under close guard, they hadn't been mistreated, and she felt it would do them good to get out into the fresh air and get some exercise. And as far as the captured soldiers went, they seemed to be dealing with the forced labor quite well. Some even did their work with enthusiasm, probably thankful to be out of the makeshift prison barn.

The Captain of the Guard was off to the side supervising, and conferring with Ruth. When they spied Marissa, they both hurried toward her, obvious relief on their faces.

"Majesty," Roland said stiffly, and bowed deeply.

"Majesty," Ruth said at the same time, and curtsied.

"It is good to see you home safely," Roland said, still showing signs of the annoyance he felt at Marissa for risking her life.

"The preparations are coming along nicely," Ruth offered. "I believe you will be pleased."

"I'm sure I will be," Marissa said. "But first, I need to freshen up. Please have a bath drawn in my chambers."

"Right away, Majesty," Ruth said, and hurried off.

Roland caught Selene's reins as Marissa dismounted, and said, "I am glad you are safe and well, Majesty. However, I must restate my concerns. You cannot let the dragon lure you into a situation from which you may not escape. You must put your own safety first for the good of the kingdom."

She smiled at him, knowing he was partly correct. "Roland," she said gently, "I know I need to protect myself. But I also need to protect my subjects. Each and every citizen of Avonridge is my child, and I will do everything in my power to keep my children safe. Now, please see to it that Selene is well cared for, she performed magnificently today." And she patted the mare's neck lovingly, then walked into the castle.

TWENTY-THREE

The hot, lavender scented water was doing wonders to soothe away the muscle aches and pains, and Marissa was beginning to unwind from the hectic pace of the past two days. The excitement she felt at meeting the legendary King Arthur was hampered only slightly by Roland's complete mistrust of the High King.

Roland was sure Arthur would attack and try to steal the throne of Avonridge. He insisted he be allowed to make ready his army for just such an occasion. And so, Marissa finally agreed, realizing it was, in fact, the prudent thing to do, just in case.

Ruth was chattering on about the preparations for the feast they would hold when the High King arrived. She was gathering Marissa's gown, jewels, crown, and all the other accoutrements the queen would need for the evening, and talking continually about the cleaning of the palace, the collecting of the foods for the banquet, and all the other details that Marisa found so tedious.

Merlin was perched on the white coverlet on Marissa's bed, lazily cleaning himself. He licked his paw three or four times, then dragged it over his face, scrubbing away some imaginary offense. She vaguely marveled at the fact that, no matter what the little black and white cat sat upon, there was never any trace of black hairs left behind.

She pushed aside mundane thoughts of her faithful little companion and tried to return to the pending events, but the bath was just too soothing. Gently splashing the hot water over her shoulders and around her throat, she let go of all the responsibilities of the kingdom for a few precious minutes. Since she'd arrived at Avonridge, she'd been the center of attention.

The ladies in waiting had to be shushed away like so many barnyard chickens, just so she could have a few moments' peace. The younger women were constantly pecking at her, telling her how happy they were to have a young queen finally. Gertrude was so old she didn't understand the young women.

The older ladies in waiting, the ones closer to Marissa's own age, or even older, were quiet and dour. They complained about almost everything, from the way the cook prepared the meals to the way the cleaning staff took care of the bed linens. They didn't vie for Marissa's attention in the same hectic, needy way as the younger women, but they were just as demanding when they insisted Marissa agree with their gripes and complaints.

Marissa had always thought ladies in waiting were supposed to be attendants to the queen, making sure she had everything she needed. This group of women, however, seemed to expect the queen to do for them. The older women acted as if Marissa should be grateful for their company, instead of the other way around.

She made a mental note to speak to Ruth about perhaps assigning these women to other tasks within the palace. She just didn't relish the thought of trying to pick different women for her attendants, as she was still not close enough to anyone to see their true personalities. And she supposed no one ever would show her their true personalities, since she was the queen. It would only be much later, after they were deeply ensconced within palace life that she would find out their little idiosyncrasies.

The water was gradually cooling, but Marissa continued to soak, fully enjoying the lavender scented solitude. Ruth was still talking about this or that, but she seemed to realize that Marissa was no longer listening. She kept up the one-sided conversation, but it was more directed at herself.

Marissa remembered the joyous reunion of the little boy Darvin with his father. When the queen and her group approached the cluster of six or eight rough wooden huts that comprised their little village, Darvin came running from his home shouting, "Father, Father!"

The man slid from the back of the soldier's horse and stumbled toward his son. Dropping to his knees, the man wrapped his arms around the ecstatic little boy and smothered him with kisses. People then emerged from the other huts and all rushed to welcome home the lucky man who had escaped being a dragon's dinner.

On the ride to the village, Marissa had asked the man, Arvin, about his little boy. She learned that Darvin was an only child; his mother had died giving him life. The father and son were as close as any two could be, since they had only each other. When he was herding the cattle and Balvindor snatched him away, his only thoughts were for his son, who would have no kin, no one to love him. Arvin had seen a little boy lose both is parents, and the boy was treated like an unwanted animal by the people of the village. Arvin assured Marissa that the people in his own village would never be so cruel, but he still had been terrified for the fate of his little boy.

Marissa instructed Arvin to spread the word that, should any child in any village in Avonridge lose his parents, the child is to be brought to the palace immediately. No child in Marissa's kingdom shall be unwanted ever again. For the first time on the long trek back to Arvin's village, she saw the man smile.

Marissa knew she'd done the right thing by riding to Arvin's rescue, and the obvious love between father and son only confirmed her feelings. Yes, she needed to use caution when dealing with Balvindor. He was an unpredictable creature, showing humane consideration one moment, and the egocentric selfishness of an animal the next. She couldn't quite figure him out. He seemed to want to be allies, but still provoked and challenged her.

She thought back to their first meeting. He had spoken fondly of Queen Gertrude, and when she questioned why he terrorized the people of Avonridge, he reminded her he was still the local dragon. He even seemed to be validating his actions by that simple affirmation of who and what he was. She had to keep telling herself that she was in a land that was unfamiliar and strange.

The rules of existence she had grown up with seemed to be from an alternate universe. The things she knew and understood from her childhood were different from the world she now inhabited. Had anyone in her previous life told her she'd be dealing with a fire-breathing dragon, she'd have told them to double their medication. And now, here she was, trying to figure out the best way to understand that very dragon.

Sliding deeper into the now cool bathwater, Marissa thought about her life. It had only been a few weeks since she'd accepted the throne of Avonridge, but it felt like she'd been here for years. Even though she was still discovering new aspects that confused and excited her, she felt comfortable and welcome.

She'd originally wondered if everyone would accept an outsider as Queen, but the people of Avonridge just went on with their daily existence. As long as there were no ripples to disrupt their lives, Marissa guessed they didn't really care who wore the crown. The only challenge she faced to her assuming the throne was Marcellus.

A chill ran through her as she thought about the old man. He had powers that frightened the people, including Sir Erick. Yet, Marissa was not frightened by him. She knew not to dismiss him, that he could cause mischief, yet she believed that was the extent of it. She felt certain he couldn't cause any real damage if she remained vigilant.

Another image came unbidden to her mind. Forsythe now invaded her thoughts. She remembered the first time she'd met him. He was charming at first, then arrogant. She hated arrogant men, yet he was somehow intriguing. Everyone in the kingdom was in awe of her, yet he treated her just like any other woman to be conquered. That was a refreshing attitude.

She had grown tired of the people she'd tried to hold conversations with. They were either too afraid to speak their minds, choosing each word with painful deliberation, or they were fawning all over her. The ladies in waiting were probably the only ones in the kingdom who were more true to their real selves, and they aggravated Marissa beyond tolerance. The younger ones were honest in their desire to be her favorite, and the older ones were honest in their indifference. She truly didn't know which was more annoying, especially when they were supposed to provide comfort and refuge from the duties of the palace.

Ruth was the only one that truly made Marissa feel comfortable. She went about her chores without complaint, even singing much of the time. She was there whenever Marissa needed or wanted her, and made absolutely no judgments about her decisions. She would even converse with Marissa on almost an equal level, but was still mindful of her station.

"Majesty," Ruth called softly, rousing her from her thoughts. "It is time to get dressed."

"Mmm, I suppose," she answered, not wanting to leave the safe cocoon of her bath. Reluctantly, she rose from the water and Ruth wrapped a huge, soft cotton towel around her. She hugged the towel to her as goose bumps erupted all over her body from the cool palace air. She shivered slightly, but actually enjoyed the sensation on her skin. Stepping out of the tub, she let Ruth lead her to the dressing table. As she sat down, still wrapped in the towel, Ruth picked up the golden hairbrush and began brushing out Marissa's hair.

After her hair was brushed, Ruth helped Marissa step into the gleaming white gown she would wear for the occasion of meeting King Arthur. It was the same one that Queen Gertrude had worn for Marissa's arrival banquet, and fit Marissa as if it had been made for her. The neckline was modestly low, with glittering diamonds encrusted around it. The tight bodice was shimmery stain, but with diamonds enhancing its glitter. The full skirt billowed softly to the floor in a luminescent flow that made Marissa think of a regal waterfall.

She silently marveled at the fact that she and Queen Gertrude seemed exactly the same size, all the clothing in Gertrude's closet fit her perfectly, and she again felt like she was in the midst of fulfilling her destiny.

King Arthur was due to arrive within a few hours, scouts had returned with details of his progress to Avonridge, and Marissa was beginning to feel excited all over again. She had read stories of Arthur and Guinevere and Camelot, and always thought she belonged in that time. Well, she had been correct, because she was indeed in that time now.

Ruth placed the emerald pendant around Marissa's neck, then attached the emerald earrings. A deep green fire danced through the huge stones, seeming to bring them to life with a warmth and vibrancy that matched the excitement in Marissa's breast. She stared at the gems, appreciating the beautiful contrast they offered to the white gown, and again thought about the impending meeting.

She wondered if Merlin the Magician would be with him, or Guinevere his wife. She didn't have very many details about him from this time, such as; how old he is, where is Camelot located, how many kingdoms has he united? But she could learn all that in due course.

Suddenly, a loud knock sounded on her door jolting her back from her thoughts. Ruth quickly went to the door, opened it and revealed Roland standing in the hallway. He looked troubled, even anxious, and Marissa waved him into the room.

"Majesty," he said, somewhat breathlessly, "I need to speak to you before it is too late." He was agitated to the point of distraction, looking around the room almost as if he expected a ghost to leap out at him.

Marissa had never seen him this flustered. As Captain of the Guard, he should be composed at all times, ready to handle any unexpected event with well-trained thoughts and actions. She gently grasped his arm and said, "Roland, calm yourself and tell me what you need to."

He nodded quickly, glanced down at the floor for a moment to compose himself, and blurted out, "King Arthur will attack Avonridge!"

Marissa sucked in her breath and drew back as if she had been slapped. She shook her head in disbelief. "No," she quietly stated. "No. I refuse to believe that."

Roland nodded emphatically, eyes huge as saucers, and repeated, "King Arthur will attack Avonridge. I am certain of it, Majesty."

Still refusing to believe, she said, "But he has no cause to. There is absolutely no valid reason for him to wage war on Avonridge. Why would he do such a heinous thing?"

"He wants to rule all of Britannia, Majesty. He is a usurper and an upstart, and he will take by force what is not readily offered."

"But that doesn't make sense, Roland," Marissa argued. "He hasn't even spoken to me about allegiance to him. Why would he assume I would refuse?"

At this, Roland faltered. "I…I do not know, Majesty," he said, his voice unsure. "All I know is that he is a usurper and an upstart, and he will take by force what is not readily offered."

"That may be so for kings that may oppose him," she said. "I understand his need to unite the kingdoms and stop the wars. But he has no cause to jump to conclusions. Everything I've read about King Arthur suggests a strong but gentle man, a man most concerned with his people. Why would he attack when he doesn't know the intentions of the people?"

Again, Roland seemed confused. He shook his head, and in a quiet, unsure voice, said once more, "He is a usurper and an upstart, and he will take by force what is not readily offered."

Marissa paused and looked at her Captain of the Guard. He had again repeated that statement verbatim, almost as if rehearsed. Suspicion began to creep into her heart, and she turned to pace the floor, trying to think through this unexpected turn of events. She paused to look at Roland, who seemed as lost as a child. She knew this was not the Captain of the Guard that she had come to admire and respect. This man was anxious and frightened, and as far as Marissa could tell, had no real reason to be.

"Roland," she said softly, "tell me something."

"Yes, Majesty," he said obediently. "Anything."

"Have you been talking to anyone about the awaited arrival of King Arthur?"

"Oh, everyone in the kingdom is discussing it. It is the topic of every conversation."

"Of course," she smiled. "It would be. But what I want to know is, have you been approached by anyone in particular about the visit? Has anyone come to you that normally may not have cause to speak with you? Someone that may normally try to stay out of your way."

His brow knitted together as he thought about her question. He slowly shook his head, as if trying to remember a fleeting moment. Finally, he looked up at her with surprise and anger in his eyes. "Yes, Majesty," he said. "Marcellus stopped me just this morning. He said he wanted to speak to me, but I dismissed him, telling him to stay away from me."

"And did he?" she asked.

"I believe so," he said, again unsure. "I do not remember any conversation."

"Do you remember how long the whole incident took?" she asked.

He again shook his head, trying to remember. "No. I thought it was only a matter of moments, but…" he paused.

"Please," she gently prodded.

"The sun," he said shortly.

"The sun?"

"Yes, Majesty. The sun had traveled more than an hour's distance from when he stopped me to when he left." The realization struck Roland with a hammer's force. Now it was his turn to pace. His boots came down on the polished stone floor of Marissa's chambers with angry thuds.

"Trickery!" he snorted. "The old man clouded my judgment and tricked me." His anger seethed for a few moments, then Marissa saw his expression change. His eyes grew large with dread and shame. "Majesty," he breathed. "I have failed you. I have allowed Marcellus' trickery to sway my judgment."

"Roland," she said forcefully, "do not berate yourself. We both know he's had much practice in his trickery. You simply need to be more vigilant when it comes to that old man. Now, tell me all you remember about that encounter."

He still stood somewhat stoop-shouldered, but related what he could. "I saw him approach me from between two buildings. I remember the sun was just cresting the building to the east. I dismissed him, bidding him leave me. I thought he did so, but was curious about the fact that the sun was now cresting the building to the west. I thought it had only been seconds, but when he left me, I had the overwhelming feeling that King Arthur was marching on Avonridge. I knew I must warn you so we could stand against him. Now, I feel foolish." His shoulders drooped even more.

Marissa again touched his arm. "Roland, there is no need to feel ashamed. Marcellus enjoys causing trouble, and he seems very good at it," she said ruefully.

He nodded dutifully, but still looked miserable.

"And remember," she continued, "we don't know for certain what King Arthur has in mind. As you pointed out, we must be prepared for anything. So, gather the soldiers. We will have them assembled as a welcome, but they will be prepared to defend Avonridge if the need arises."

She saw his anguish diminish slightly. He stood up straighter, and quietly said, "Yes, Majesty, the troops will be prepared and ready. I have already alerted them, so we will be in formation to welcome the High King."

Marissa heard some trepidation in his voice. He obviously still did not fully trust King Arthur, but was at least willing to wait for a legitimate cause before he started any military action.

TWENTY-FOUR

The crest of the Pendragon flew high above the awe-inspiring army that marched toward the lowered drawbridge of Avonridge. Marissa stood on the bridge, Sir Erick several paces behind her, vigilant as always, waiting to welcome the High King. Merlin sat majestically at her heel, purring loudly. Her heart was racing with excitement and more than a little fear. What if Roland had been correct? What if Arthur was marching to seize Avonridge without benefit of meeting with Marissa to see where she stood? His army was sizable, and she knew from the legends and stories she'd read about him that he was an accomplished leader. Would her army be able to stand against Arthur if he attacked?

Approximately one quarter of her troops stood outside the gate of the kingdom, flanking the drawbridge on either side, and lined up three deep along the moat. The remainder of the army waited inside the fortress walls with Roland. If the unthinkable happened and fighting began, the majority of her soldiers would be inside to defend the castle. Whatever enemy soldiers managed to get inside before the gate was secured would be met by fierce opposition. Even though she was sure in her head that Arthur wouldn't be so reckless as to attack without need, her heart still pounded in her chest.

Still mulling over the slight, but distant possibility of attack, she felt a ripple in the air around her. The pit of her stomach developed a knot of ice, and a chill ran up her spine. She recognized the feeling, and turned to see Marcellus hobbling toward her on the drawbridge. "Great," she muttered to herself. "Just what I need now," and Merlin echoed her sentiment with a menacing hiss. She looked at Sir Erick, who was watching intently. She was relieved to see the Black Knight knew Marcellus was here. It seemed as if the old magician could hide at will from everyone's awareness but Marissa's, but now didn't seem to want to do so.

The old man stopped a few feet from her and gave a half-hearted bow of his head. "Majesty," he rasped, and she could hear the sarcasm he put into the title. "I see you are prepared to meet the High King."

"I don't have time for you just now, Marcellus," she snapped, and Merlin growled softly, arching his back and puffing up his coat.

He bristled visibly at her rebuff, but persisted anyway. "I can see you are otherwise engaged," he said, his tone slightly mocking. "However, I do not believe you have sufficiently thought through the manner in which you meet the High King."

Not wanting to rise to the old man's bait, but unable to think of an easier way to be rid of him, she decided to allow him his say. "Alright, what is it you think is amiss? I have my army in welcome formation. Just what is wrong with their 'manner'?"

He lowered his head to disguise it, but she could see his smirk. "It is only that your soldiers are fully armed. The High King will take that as an insult. He will most definitely believe you are planning an offensive strike. Why else would you have them ready for war?"

A stab of doubt pierced her heart when she heard the reasoning of his words. But then she remembered his tricks. He had convinced Roland that Arthur would attack Avonridge. Now he was trying to convince Marissa that Arthur would feel threatened. He obviously wanted an armed confrontation between Avonridge and King Arthur, and he didn't seem to care from which side it came.

She drew herself up to her full, if somewhat diminutive, height of five feet five inches, and loudly stated, "Marcellus, your advice is appreciated, but completely unnecessary," and she made sure to match the sarcasm he had exhibited. She was gratified to see it work, as he grew angry.

"Majesty," he said, the word positively dripping with contempt, "you would do well to consider my words. You do not want to make an enemy of the new High King."

"Marcellus, you confuse me," she declared. "First you convince my Captain of the Guard that the High King will attack Avonridge. Now you want me to believe he will feel threatened by my show of respect. Which is it, old man? Do we need to defend ourselves, or does the High King?" Her voice had become like ice and her stare bored directly into his eyes. He tried to stand defiantly against her, but he faltered and took a half step backward and lowered his gaze.

"You underestimate many things, Majesty," he said quietly, and she could hear pure hatred seething just below the surface of his façade. "I pity you when it all comes crashing down around you." And he turned away from her.

She watched him hobble away, and caught the Black Knight's eye. She could swear he was smiling beneath his leather mask. She gave him a quick nod of acknowledgement, then returned her attention to the approach of Arthur. Unfortunately, she continued mulling over the idea that Arthur would indeed find her armed soldiers threatening. She tried to dismiss the thought, but it persisted. Finally, she mentally stamped a foot and commanded herself, "Stop that right now! It is simply that old magician trying to influence me. I'll have none of it, and I know everything will be just fine." Abruptly the doubt vanished, and she could concentrate on the matter at hand.

The High King's procession was moving smoothly across the field toward Castle Avonridge. There were three riders in front, and approximately a thousand mounted soldiers in formation behind. Of the three in front, one rode in the center, and the two flanking him carried the banners. One Marissa recognized as the Pendragon Crest, and the other was the banner for Britannia. Since this army consisted of far less soldiers than her scouts had conveyed, she had sent them back to locate the rest. They were camped a few miles from Avonridge, and obviously not readying for an assault. This small army was most likely insurance against an attack from Avonridge. Not enough soldiers to actually capture the kingdom, there were just enough to defend their king long enough for him to escape and return with the remainder of his forces. That fact brought great relief to Marissa after all the worrying about an unwanted battle.

The procession was finally close enough for Marissa to decipher the lead rider. She had no actual knowledge of what King Arthur looked like, but was certain that this was he. She held her breath, unaware of everything except the approach of the High King, and her aggravatingly heavy crown and royal cape. The crown felt like it weighed a ton and made her head throb, and the cape made her sweat, even in the cool autumn air, but she would meet the High King in all her royal glory.

When the procession reached perhaps two football fields length from the drawbridge, Arthur raised his hand and stopped the advance. Marissa's heart skipped a beat when doubt again stabbed at her. But Arthur dismounted his horse, and the soldiers all followed suit. Recognizing the act as a gesture of greeting, she smiled and finally began to breathe again. With none of his soldiers mounted, it would be impossible to wage a surprise attack, and so he was in fact showing his great desire for a peaceful meeting and fruitful negotiations.

The two men carrying the banners remained astride their mounts, and Arthur handed his horse's reins to one. He then began walking toward Marissa alone. Not willing to show the slightest disrespect in making him bridge the distance himself, she also began walking. Realizing that the High King of Britannia had made himself completely vulnerable in his assurance of her loyalty, all the fears that had plagued her and Roland for the past hours simply vanished.

As they approached each other, Marissa studied the High King. He seemed so young, appearing not yet twenty. He was not a large man, only standing about five feet and ten, and not very heavy. But he carried himself with an air of regal authority. Marissa knew from the Arthurian legend that he'd begun life as a page, totally unaware of his royal destiny, and she could see humility in his eyes. Instead of the arrogance one would expect from the High King of all Britannia, she saw warmth and openness. That, mixed with his self-assured posture, created a man that could truly lead nations.

He walked with an easy stride, as if he had not a care in the world. Without the headpiece of his armor, his short dark hair framed a boy's face. He had a friendly expression, almost a smile, and Marissa instantly liked this man. His hard iron armor made soft but distinct clanking sounds as he walked, and she absently compared it to her own golden armor that was whisper quiet. Knowing in her heart she had made the correct decision in offering her allegiance to this man, she fairly skipped in her eagerness to meet him.

The two sovereigns stopped three paces from each other, and together showed their mutual respect. Arthur bowed his head in greeting, and Marissa curtsied deeply, her satin skirt billowing out around her in regal elegance. Merlin stood next to his queen, little black head held high, and meowed loudly in greeting. Arthur looked at the little creature, and gave a short, easy laugh.

"My lord," Marissa said, "all of Avonridge welcomes you. Even my little feline friend," and she rose, also laughing at the cat's forwardness.

Arthur's smile was genuine, and his eyes danced as he said, "I thank you, Queen Marissa."

TWENTY-FIVE

Yet another celebration was underway at Castle Avonridge. The great feast to welcome the High King was going well, and Marissa felt gratified in the festivities. The castle had been cleaned and was absolutely gleaming, the food was plentiful and delicious, and the music that filled the banquet hall was light and cheerful. Ruth had seen to the setting up of a huge banquet table, with two high-backed woven chairs set at the center. Marissa wondered if Arthur's should be just a little more elaborate than hers in respect to his position, but Ruth assured her that, as sovereign of Avonridge, she should be on an equal level.

On either side of the two places of honor sat the who's who list of Avonridge. Sir Erick sat at Marissa's left hand, always her faithful protector, and to his left sat Roland. The Captain of the Guard balked at being included in the royal line-up, but Marissa had insisted. Beside Roland sat some of the members of the court in higher standing. Lords and Ladies were leaning forward trying to catch the attention of the guest of honor, and he was politely nodding and answering an occasional question here and there.

To Arthur's right sat the true Merlin, the magician and mentor to the High King. Marissa had had little time to study the legendary figure, and was anxious to know all about his magic. She knew from the stories she'd read that he had considerable power when it came to influencing the minds of his enemies, after all, he'd arranged for the conception of Arthur. She also kept in the back of her mind that he may try to influence her if he felt she may not offer Arthur the allegiance he desired. Well, that was inconsequential, since she fully intended to pledge her kingdom to his service. She also wanted the wars between the kingdoms to stop and the lands to live in peace, and Arthur was the only one that could accomplish that great feat.

Using the distractions of the banquet to study the young king, she noticed that, even though he was slight in build, he was powerful. He had shed his heavy armor and sat comfortably in his royal garb. He wore a long-sleeved purple shirt made of silk, with gold piping along the seams, and as he moved, muscles rippled beneath the smooth cloth. A simple gold band encircled his head, and black cotton breeches fit nicely over his well-toned thighs, with the legs tucked inside black leather boots. He had an outgoing manner; laughing easily and making Marissa feel on an equal level with this, the most powerful man in Britannia.

His slight build belied his well-muscled strength. Being used to the 'Mr. Universe' sized bodies of pro-wrestlers and the like, Marissa found his quiet strength refreshing. He was obviously well suited for battle, but was not overly built to the point of excess. She liked the unassuming nature of this man, who possessed an air of total control and self-assurance. His ease in making conversation was quite enjoyable, and Marissa fully appreciated the genuine manner in which he accepted the good wishes offered by, it seemed, everyone in the kingdom. As the evening wore on, members of the court continually flowed up to the table, obviously wanting to get a glimpse of, and to have the chance to speak with, the High King. It was a once-in-a-lifetime opportunity, and Marissa didn't really begrudge them their desire to take it.

However, when she saw Forsythe approaching, an unexpected dread descended over her heart. She thought the feeling was unnecessary and unreasonable, but that only deepened it. Her instincts were shouting something at her, but she couldn't fathom exactly what. She knew Forsythe was an opportunist, but couldn't understand why she was so vehemently opposed to him speaking to Arthur. She wanted to scoot him away, but didn't want to call attention to the fact that she was doing it. Not knowing an unobtrusive way to be rid of him, she had to wait and see what he had in mind.

When Forsythe approached the High King, he made a grand bow at the waist, fairly doubling over and swinging his arm out with a flourish. "Most Royal Majesty," he gushed. "I bid you welcome to Avonridge."

Arthur glanced sideways at Marissa, and she was sure she saw caution in his eyes. "I thank you, sir, for your welcome," he said. "However, your sovereign, Queen Marissa has already extended the welcome of her kingdom."

Marissa's heart skipped a beat when she heard the rebuff in Arthur's voice. He was clearly putting Forsythe in his place. Unfortunately, the arrogant man either didn't realize it, or chose to ignore it. He rose from his bow smiling.

"Of course, Sire. The good queen knows how to perform the duties of the royal household. She is certainly accomplished when it comes to entertaining guests."

The insult was not lost on Marissa, nor, she could see, on Arthur. Forsythe had made it clear that he felt Marissa should stick to the running of the household, instead of the running of the kingdom.

Arthur opened his mouth to speak, but Marissa beat him to it. "Forsythe," she said, her words dripping with obvious false honey, "how gratifying it is to hear your compliments. Knowing that my efforts at making the High King feel welcome meet with your approval means a great deal to me. Now, if you will excuse us, we were discussing some very important matters of state." Her face had become a stone mask with a slight smile fixed in place.

Again, he either didn't hear the warning in her words, or decided to ignore it—Marissa was quite certain it was the latter, as Forsythe was not a stupid man—and continued to smile at the High King. "Of course, I must be going. I have some very important business to attend to. But I would greatly appreciate the opportunity to have a few words with you in the near future, Sire. A few words in private about a personal matter that is quite dear to me."

Arthur nodded slightly, his face a mask of suspicion. "Of course," he said. "Simply ask your queen to arrange a meeting, and if I am available, I will be happy to oblige."

Forsythe smiled, bowed again, and said, "Thank you, Sire," and he walked away. The fact that he had not greeted Marissa when he arrived, nor acknowledged her when he left, made a white-hot flame of anger burn in her breast.

She voiced that to Arthur, adding, "How dare he!"

Arthur nodded in agreement. "Yes, he doesn't understand the rules of protocol."

"Oh, he understands them," she spat. "He just doesn't believe they apply to him."

"I told him to arrange the meeting through you, hoping he would change his mind."

"He won't," she stated flatly. "And he won't come to me, either. He'll manage to get an audience with you without my knowledge."

"Then I will simply refuse to see him," Arthur promised.

She thought about that for a moment, then shook her head. "No. If you don't mind, Sire, go ahead with the meeting. Not knowing what he has planned is an option I don't want to exercise. I haven't trusted that man since the day I arrived. I want to know what he has in mind."

"Of course," he agreed. "And I'll relay that information to you immediately."

She felt a little more at ease knowing he was on her side. She allowed Arthur to be fawned over by the members of the court as she tried to figure out Forsythe. She had entertained thoughts of keeping company with him, remembering how charming he was when he'd presented her with the rose. But then he'd reverted back to the arrogant ass he was at her arrival banquet when he'd claimed he would marry her. Now he obviously wanted to ingratiate himself with the High King. She off-handedly wondered if he would try to persuade Arthur to force Marissa to marry him. No, that was too ridiculous to even consider. And she felt sure Arthur wouldn't go along with something like that. Arthur didn't seem to like Forsythe any more than Marissa did.

The banquet lasted until well after midnight, and Marissa was seeing to the cleaning of the hall when Ruth came up to her. "Majesty," she scolded. "I will make sure the staff does its job. You get your rest," and she gently pushed Marissa toward the hallway that led to her chambers.

Grateful for the chance to finally rest from the excitement of the past several days, she obediently started meandering down the hallway. She was almost to her door when a voice called out to her.

"Majesty, if I may have a moment." She turned to see Arthur's mentor, Merlin the Wizard, walking toward her. He was a large man, perhaps an inch or two shorter than Sir Erick, and he walked with a quick, self-assured step. He looked old, with long white hair and beard that reminded Marissa of Marcellus. But where Marcellus had an air of malignancy, Merlin proffered a feeling of honesty, even benevolence.

The age on his face and the youth in his stride seemed at odds. He looked as old as time, but appeared young and spry. His rich charcoal gray robes billowed out around him as he walked, giving him the impression of flight. When he reached her, even though he'd been hurrying, he was not at all winded, and breathed slowly, like a well-trained athlete. He stopped a few feet from her, and gave a quick bow.

"Majesty, I know it is late, but I was hoping we could visit for a few moments."

Wanting desperately to get to know this legend of a man, she quickly said, "Of course, of course. Please come in," and she led him into her sitting room. Across the room was the small table with the two chairs, and perched majestically atop the table was the little black-and-white cat, Merlin. He sat still as a statue, but gave a loud "Meow," when the queen and the magician entered.

"Ah, my little namesake," Merlin stated, and marched straight to the cat. He stood up and commenced purring like a motor when the big man began stroking his fur. "I haven't seen you in ages, my friend. I trust you've been well?"

The cat meowed again, and rubbed his head against the magician's hand, fully enjoying the attention. Marissa watched the exchange in rapt silence, again completely confounded by the fact that everyone, absolutely everyone, in this kingdom seemed to have a personal relationship with her little black-and-white ball of fluff.

"Excuse me, Merlin," she said, "but did you call him your namesake?"

"Indeed, I did, Majesty," he replied, still petting the cat. "This little creature and I have been friends forever, it seems. When Queen Alice wanted a guardian for the heir to the throne, she asked for my advice. I, of course, chose this little spitfire, and as a tribute, she named him Merlin." The big man was using both hands to playfully rub and pet the cat, and both seemed lost in the enjoyment of it.

"Queen Alice?" Marissa said. "Do you mean you knew Queen Alice? And Merlin's been around since her time? Please tell me how all that came about."

"Ah, Majesty," he said, finally turning away from the cat, "that is a story for another day. Right now, I wish to get to know you. May we sit and discuss the wonderful things that are happening right now?"

The cat jumped off the table and walked into Marissa's bedchamber, throwing a "Meow," over his shoulder as he left the room.

"Yes, good night my friend," the old magician called. "We'll meet again soon." Then he turned back to Marissa, and said, "Please, Majesty. Sit with me and talk."

She absently walked to the table and sat down, still more than a little confused at the information she had just been given. By the Histories, Queen Alice had reigned two thousand years ago. Was this man claiming to be that old? And was he also saying that her faithful little cat was also that old? More mysteries, more questions. Just when she had begun to feel like she was getting comfortable with being a queen, and coming to know Avonridge, something else would spring up to confuse and challenge her.

Merlin took the seat across the table from her and said, "I am grateful to see that you have accepted Arthur as High King."

Still lost in thought, her only reply was, "Hmmm."

"You have accepted Arthur, have you not?" Merlin persisted, somewhat uneasy.

She looked at him and forced her mind back to the present. "What? Oh, yes, yes of course. I am well aware that King Arthur is the Pendragon."

"Good," he said, and relaxed a bit. "It is imperative that we unite the kingdoms. After all, the wars have been going on for generations. One would think the kings would have grown tired of battle by now. Alas, it is the way of man to fight, is it not, Majesty?"

She smiled at his invitation to debate the war-loving ways of men, and chose instead to ask a question of her own. "Tell me, Merlin, how old is Arthur? Has he come into his own yet?"

The old man smiled and leaned back in his chair. "Arthur is still a boy, as you have seen. He'll be nineteen in three months' time, but he's been an adept pupil. His father's cunning and strength, and his mother's wisdom and gentleness have combined to create a truly magnificent leader. His men would follow him into the fires of Hell if he desired it."

"Yes, I can see that," she said. "And so young. He truly is destined for greatness."

"But what of you, Majesty," he said. "Do you feel the same loyalty to the High King as his soldiers? Can Arthur depend upon your allegiance?"

"Oh, Merlin," she said, "I, too, want the wars to stop. I've just finished burying my own dead soldiers over a stupid and senseless battle. Arthur is the only man capable of putting a stop to the fighting. As long as he is High King and seeks unity and peace, my kingdom is at his service."

He seemed satisfied, and nodded. "Good. So far it has not been going as smoothly as we'd hoped. So many of the lesser kings seem to be waiting for someone else to be the first to swear allegiance. They're either afraid to further anger the other kings, or simply do not want peace. As the only kingdom in Britannia to have been able to evade the wars for so long, if you can make it known that you support Arthur, it should make the other kings consider the option. Having you as an ally will help the young king gain credence."

"I am honored," she said. "I had no idea my support could mean so much."

"Indeed, Majesty," he assured her. "Avonridge has endured in peace for many years. All the neighboring kingdoms acknowledge the wisdom and strength that sits upon the throne. They would indeed follow your example."

"But Merlin," she argued, "they all know I've only just ascended the throne. They all know I'm a novice at this."

"But they also know you've been chosen by Queen Gertrude. They will respect that fact and give you the same reverence they gave her."

"All except Sorens of Southcross," she muttered.

"Ah, Majesty, King Sorens should be no more trouble to you. He is not long for this earth."

She looked at him with interest, and asked, "You can see that? He will die soon?"

"Soon enough," Merlin stated.

"Tell me more," she urged. "Will he be killed, or die of some disease?"

"Details elude me," he said, waving his hand. "But his time is drawing to an end."

Marissa nodded, feeling relieved that her first, and hopefully only, enemy would soon be able to bother her no more, but she also felt guilty about wishing for his death.

Merlin seemed to sense her feelings, and said, "Majesty, do not trouble yourself over something that is beyond your control. You will not determine the circumstances of Sorens' death, he will."

She accepted that, and changed the subject. "So, great wizard," she said, smiling, "please tell me how you know my furry little friend. He seems to know everyone of importance here, and so he's become quite a mystery to me. I was under the impression for fifteen years that he was just a cat."

Merlin smiled, and said, "Impressions are almost always deceiving. But I am sure you're tired. You need to rest up for the activities of the coming days." And he stood up and started for the door.

"Oh, no, please don't go yet," she protested, jumping up and following him. "There are so many questions, so many things I need to know. Please stay and talk with me."

"There is plenty of time for that, Majesty," he said gently. "Right now, you need to rest. We will talk again soon."

"That's what Queen Gertrude said," she remarked miserably.

"There was a reason for that, as well," he assured her. "But fear not, my time has not yet come." And with his robes billowing out around him, he marched through the door and out of view.

TWENTY-SIX

The tournaments to honor King Arthur and welcome his army were going splendidly. Box seats were built and decorated just for the occasion. A platform had been erected with three wooden walls and roof built around it to try to keep the windborne dust and dirt to a minimum. Royal purple cloth was draped across the front of the roof to acknowledge the queen's presence, and soft purple cushions were thrown onto the wooden seats for comfort.

Marissa sat with Merlin the Magician on her left, and Merlin the Cat in her lap. Try as she might to get the old man to tell her any more about the little cat's history, he adeptly evaded all her questions, steering her attention and the conversation back to the jousting field. Her annoyance at being put off was short-lived, however, when the combatants all filed out onto the grounds.

Magnificent to behold, the knights of both Marissa's kingdom and Arthur's entourage were dressed in gleaming armor. They sat atop their steeds filled with pride and elegance, fully enjoying the admiration of all who had come to witness the contests. Their huge warhorses were equally bedecked, with leather saddles oiled to rich finishes, and bright red plumes affixed to their bridles. They snorted and pranced in anticipation of the games, and lent an air of excitement to the watching crowd.

Marissa was disappointed that Arthur was not watching with her, but he had preferred to join the games and joust with his men. She supposed she understood him wanting to show off, but would have liked to talk to him more about his plans. At the moment, it seemed she knew more of his future than he did, having avidly read everything she could about King Arthur and Camelot before she came to Avonridge.

Oh well, plenty of time to talk later. Right now, the knights were parading around the edge of the joisting field, allowing all the spectators to "ooh" and "aah" over them. Ruth had joined Marissa and was seated to her right, cheeks flushed pink with excitement as she leaned forward, trying to catch every glimpse of the magnificent show.

"Oh, Majesty," she breathed, "we haven't had games for so long. This will be just wonderful to watch." Marissa smiled and nodded at her companion, allowing herself to join in the feeling of excitement and anticipation. Merlin the Cat stood perched on her lap, watching every minute detail of the festivities that were beginning.

As the knights paraded around, they stopped before the royal box. All of Arthur's knights, with Arthur in the lead, lined up facing Marissa, and in unison, bowed to the queen.

Then, in a clear voice, Arthur loudly declared, "Majesty, we are honored by your presence. May each of us succeed in providing enjoyment to you and your subjects." Then he led his group away, and Marissa's knights replaced them.

Also lining up before their queen, Sir Erick led this group, and he echoed King Arthur's words, stating loudly, "Majesty, we are honored by your presence. May each of us succeed in providing enjoyment to you and your subjects." Then he led the knights onto the jousting field.

An expectant hush descended upon the field, as all seemed to hold their breath in unified anticipation. Marissa could almost hear the multitude of beating hearts around her as the excitement climbed to a fever pitch, sending bursts of electricity through the crowd. The air fairly crackled with the promise of spirited entertainment.

King Arthur and Sir Erick, being the highest-ranking soldiers on the field, would begin the games. Facing each other across the expanse of the tournament arena, they prepared for the first joust. Pages brought the jousting lances to the combatants, and made final adjustments to their armor. As the horses stamped and snorted with impatience, king and knight raised their lances in salute to each other, then lowered the visors on their helmets, and began their advance.

Warhorses are large animals, with little speed, but extreme power. Their gaits were lumbering at first, then increased as they gained momentum. Beginning approximately two hundred yards away, it almost seemed as if they may never reach each other, but when they did, it was with great violence.

Both lances connected; Sir Erick's slamming into Arthur's shoulder; and Arthur's crashing into Erick's chest. The lances shattered with screeches and snaps that echoed through the arena. Animalistic grunts emanated from the combatants, as each was thrown back across their horse's rump, but neither was dislodged. As they completed their run, King Arthur and Sir Erick both remained astride their mounts.

The crowd erupted in cheers, and Marissa finally took a breath. She turned to Merlin and said, "Don't you think this is a little dangerous? Suppose the High King gets hurt?"

"Should the High King get hurt participating in games with friends," Merlin calmly stated, "he does not deserve to be High King."

Marissa chuckled at the old man, then returned her attention to the field.

King Arthur and Sir Erick were preparing for their second run, and the crowd was again quieting down. New lances were brought, and adjustments were again made to the armor. Then the run began, just as slowly as the first, and increasing with each step of the magnificent animals. Again, the warriors met with the clash of wood on metal, again lances shattered, throwing sharp shards of wood into the air, again both combatants were thrown back across their horses' rumps, and again both remained seated.

The crowd cheered, and Marissa could feel pride in her warrior knight. He was battling a legendary king, the greatest warrior of their time, and Sir Erick was performing magnificently. He was not backing down as he eagerly received the third lance, and turned for yet another run. Marissa wondered just how long both men could continue.

The third run was almost identical to the first two, except that King Arthur's foot was dislodged from his stirrup when he was thrown back, making it appear as if he would be unseated. The crowd gasped as they watched the High King wobble slightly, and then cheered loudly, chanting "Sir Erick, Sir Erick!" accentuating it by loud foot stomping.

Arthur managed to remain in the saddle, however, and a feeling of dread encircled Marissa's heart. How would the king, indeed all his troops, react if Sir Erick bested Arthur? That could cause extreme tension, if not outright conflict. Knowing that nothing good could come from Sir Erick winning this competition, she wondered if she should call it a draw and put a stop to it. However, Merlin again seemed to sense her thoughts.

"Most enjoyable game, it is not?" he said easily.

She nodded absently, silently wondering what to do. The jousters were preparing for their fourth run, both showing signs of tiring.

"It should not be long now," Merlin noted.

Marissa wanted desperately to do something, but knew there was nothing for her to do. At this moment, her place was in the royal spectator's box, and her only course of action was to watch. And in watching, she noticed Arthur sit up a little straighter, with shoulders slightly farther back. The tilt of his head was just a bit more regal, and he held the lance with an easy grace that seemed to challenge the very world around him.

Sir Erick was accepting his fourth lance, and Marissa could sense a feeling of triumph about the Black Knight. He had essentially drawn first blood, having dislodged King Arthur's foot before his own, and as a warrior, he would now be preparing to finish him off. The feeling of dread in Marissa's stomach grew and knotted until it felt like one of the stones used to build the castle. She couldn't blame the knight for wanting victory, but she didn't know King Arthur well enough to know how he would react to defeat. Could their alliance be doomed even before it had really begun?

A dull ringing began in Marissa's ears, and she held her breath as the combatants again began their run. The warhorses' hooves pounded the ground as they again made their way down the jousting field. The rest of the world disappeared from view, as Marissa watched what could be a disastrous game. No sound penetrated her hearing except that of the drumbeats of the charging beasts, and a sudden spell of lightheadedness made her grip the wooden railing before her to keep herself sitting upright.

The horses charged toward each other, gaining speed, but to Marissa the scene unfolded in slow motion. It felt like hours for the combatants to close the distance between them. But, finally, in the instant before they met, both riders brought down their lances to connect, and Marissa noticed Arthur lean forward ever so slightly.

As the lances made contact, Sir Erick's glanced off Arthur's shoulder. Arthur's lance, however, hit home perfectly. Slamming into the center of his chest, the force of the blow lifted the Black Knight cleanly out of his saddle. He seemed to hang suspended in the air for several seconds, then fell to the ground in a crash of clanging metal, sprawling flat on his back.

Marissa jumped to her feet, and the entire audience followed suit. Silence hung like a blanket in the air, as the Black Knight lay motionless for an eternity. Just as Marissa decided she should rush to check her friend, he rolled onto his stomach and lifted himself to his knees. The crowd erupted in cheers and applause, and Marissa's heart made its way down from her throat back into her chest where it belonged. She watched as Sir Erick stood up, and was relieved to see he seemed no worse for wear.

The crowd continued to applaud, and, as fickle crowds are known to do, transferred their appreciation to the High King. The chant now consisted of, "Arthur, Arthur!" With Merlin sitting beside her, a quiet look of satisfaction on his wrinkled old face, Marissa only smiled.

The hand-to-hand combat exhibition was held the following day. This was Marissa's strong suit, and she wished she could be a part of it, not just a spectator. After all, she had already bested a king in swordplay, and would love to show Arthur her ability. Unfortunately, again she was relegated to the spectator's box.

The main event today would not be a battle between King Arthur and Sir Erick, King Arthur was in the royal spectator's box with Marissa and Merlin. Marissa wished she could be on the battlefield participating in the activities, but realized it was not prudent. Her role was mainly to be host to Arthur and his troops, and she would do so graciously.

After the opening formalities of the knights addressing the king and queen, there were several staged battles. Combatants fought with gleaming swords, attacking and repelling as if it were true warfare. Maces were used next, thick sticks holding menacing looking metal balls with wicked looking spikes anchored by a heavy chain. They crashed onto their opponents' shields with such force that Marissa thought sure the shields would shatter. The crowd was once again enthralled by the festivities. They cheered for their favorites, and booed eagerly for the others.

About midway through the day, after a half-hour intermission, a new battle was staged. The main combatant was an evil-looking knight dressed in black. His appearance was completely different, however, from that of Sir Erick, as this knight was obviously the 'bad guy'. Where Sir Erick looked mysterious, but more defensive than threatening, this knight's demeanor was harsh and wicked. The helm he wore was shaped like a vulture's head, sharp and devilish, completely covering his identity. His black horse was bedecked in black leather, with the same vulture's head attached to his bridle, and black feathers were affixed all along its reins. This knight represented all that was unholy in the world.

When he entered the playing field, the crowd roared. Some obediently booed the bad guy, and others cheered, knowing he would ultimately be defeated but would first put on a good show. The knight rode around the perimeter of the arena, giving all spectators a good look at his wickedness, and growled theatrically. The roar grew in volume, appreciative applause resounding throughout the cool midday air. He then made his way to the center of the arena, and stopped. His mount reared, pawing the air in defiance.

As the evil knight waited, a knight of the kingdom of Avonridge entered the arena. He was dressed in the usual chainmail, shiny and pure looking. His horse was a light chestnut, with red plumes draped from its reins. This knight represented all that was good and right. The crowd welcomed him, cheering loudly. He rode to the center of the arena where the dark knight waited, and bowed to his opponent. The dark knight didn't return the salute, but stared at him in arrogance. This made the crowd boo loudly again. Some threw their fists in the air to signify their total disdain of the evil knight.

The two combatants turned their horses to face the royal spectator's box. They both dismounted, and pages came running to take their horses off the field. The two knights bowed to Queen Marissa and King Arthur, but where the knight of Avonridge bowed deeply and reverently, the dark knight bowed slightly, as if arrogance prevented him from fully acknowledging the royal duo. Again, the crowd voiced its contempt, roaring and jeering.

Suddenly, the evil knight drew his sword and attacked. The knight of Avonridge was not completely caught off guard, and so managed to sidestep the assault long enough to draw his own weapon. The battle was ferocious, swords arcing through the air, sunlight bouncing off the polished metal with blinding fierceness. The clang of sword upon sword reverberated through the stands, as the crowd seemed evenly split with its loyalty. Both knight of Avonridge and evil knight were cheered and encouraged.

The prowess of the evil knight was quickly evident, however, when the knight of Avonridge was pushed backward with such relentless force, that he finally lost his footing, and the evil knight stood above him with sword tip pointed at his throat. Acknowledging defeat, the knight of Avonridge lay his own sword on the ground in resignation. Having proven his exceptional ability, the evil knight made this battle short-lived.

The crowd roared. The evil knight only slightly acknowledged them, turning to face the stands momentarily, then quickly turning away. His arrogance further incited the crowd, and they played along, fully enjoying the spectacle.

The knight of Avonridge rose to his feet and walked from the arena. When he was out of sight, Sir Erick rode into the arena, in his usual black attire. This would be most interesting; dark knight against dark knight, yet one representing good and the other evil.

The crowd welcomed Sir Erick, cheering for their champion, and he rode his horse around the arena, nodding to the crowd in acknowledgment. When he reached the center, he dismounted and bowed to the royal spectator's box. He drew his sword from its scabbard and held it before his face, tip pointing Heavenward in a salute to his queen and High King, then sheathed it and turned to accept his shield.

The evil knight was standing off to the side in arrogant defiance, and when Sir Erick had his shield strapped to his arm, he moved forward. Sir Erick turned to face his opponent, and both dark figures stood for a moment, allowing the crowd to cheer and applaud the combatants.

In fluid motions, both knights drew their swords in unison and attacked. Shields were brought up to ward off the swords, and metal clanged against metal. Swords sliced through the air with lightning speed, only to be met by heavy shields. The two knights fought viciously, each seeming on an equal par with his opponent. One moment the evil knight drove Sir Erick back a step or two, then Sir Erick gained the upper hand and repelled the evil one, driving him backward.

"Oh, how I wish I could be on the battlefield," Marissa said, more to herself than to her companions.

"Yes," Arthur replied. "Your prowess with the sword is well known."

"It makes me feel in total control when I wield the Dragon Slayer," she acknowledged.

"I have always enjoyed hand to hand battles, they give you the chance to show off," he said.

She smiled at him, then returned her attention to the two seasoned knights doing battle.

It still seemed a battle of equals, both knights gaining, then losing ground. Suddenly, instead of a full-frontal assault, Sir Erick stepped to the side just enough to throw the evil knight off balance. The tip of his sword dropped down, and Sir Erick brought his blade down across the evil knight's, knocking it further toward the ground. Sir Erick then brought his elbow up and into the front of the evil knight's helm, driving the evil knight backward awkwardly.

Sir Erick then spun and plowed his shield into the evil knight's chest, fully unbalancing him, and he staggered several steps, but managed to remain upright. Sir Erick then brought his sword across at waist level, slamming it into the evil knight's side, and knocking him to the ground. The evil knight fell with a loud metallic clang and a dull thud, but immediately rolled to his side and was up on his knees in an instant. The crowd roared and applauded the evil knight's agility.

Sir Erick, however, did not let up his assault. He brought his sword down in an arc aimed at his opponent's throat, but the evil knight was quick and brought up his shield, effectively deflecting the blow. The evil knight tried to rise, but Sir Erick was relentless, slashing at his armor this way and that, forcing the combatant to concentrate on defending himself. Sir Erick threw all he had into defeating this evil menace, completely playing the role of defender of the realm. He dropped his shield to grip his sword with both hands, and slashed with all his might.

The evil knight was overwhelmed, and fell to the ground once again, losing his grip on his sword. He tried to bring his shield up to protect himself, but Sir Erick kicked it aside, and brought his armor-clad foot down onto it, holding it on the ground at the evil knight's side. The evil knight lay completely defenseless, and Sir Erick pointed his sword at the evil knight's throat. All movement stopped, all spectators held their breath. The evil knight lay motionless on the ground and Sir Erick stood above him in complete victory.

The crowd erupted in a roar of elation. Their champion had defeated evil, Avonridge was safe from the menace, and all was right with the world.

TWENTY-SEVEN

Marissa and Arthur walked through the wooded paradise in easy company. The conversation was light and intermittent, both feeling comfortable enough to not need forced small talk. Mention was made here and there of the beauty of the forest, the peaceful feel of the magnificent trees. The very air seemed to comfort and protect the Queen of Avonridge and the High King of all Britannia.

Arthur was clad in a leather tunic, with dark green leggings and leather knee-high boots. His only weapon was the legendary sword, Excalibur, strapped to his side. Marissa noted he seemed never to be without it. And she fully understood his attachment to the beautiful weapon. This sword was the one Arthur drew from the stone to claim the crown of the Pendragon. This sword was the one that placed Arthur's destiny at his feet. Should any danger present itself, Marissa felt completely safe within the company of Arthur and Excalibur.

The queen herself was in regal dress, wearing a gown of gleaming white. Queen Gertrude had seemed to favor the white in her attire, and Marissa now adopted that practice as well. White symbolized pureness of heart, and her heart was filled only with love for her kingdom and her people. She wore the simple gold tiara, instead of the heavy crown, preferring to use the crown only for official purposes, as it was most cumbersome and uncomfortable.

As always, her faithful little companion Merlin trotted at her heels. He never left her side, and she enjoyed the feeling of having him near. His presence was a quieting hand on the crazy ride that had become her life.

Sir Erick had wanted to accompany the two monarchs on their stroll, and was visibly distressed when Marissa insisted he remain behind. The Black Knight was never happy when his queen was out of his sight, and therefore out of his protection. So many different dangers lurked everywhere, and his primary duty was to keep her safe. Even with all her assurances of Arthur's ability to protect her, he was still quite agitated when the two walked away.

The stroll through the forest was not just a leisurely activity. Marissa wanted to get to know Arthur away from the trappings of the palace and royal duties. This young man, young enough to be her son, would be the greatest leader of this land. She wanted to know how he felt about the other lesser kings within his realm, how he planned to stop the wars and bring the kingdoms together under peace. And she wanted him to know she would support him completely.

They emerged from the woods and entered a clearing. Straight ahead lay a vast lake of clear, still water. They walked to the edge of the lake, and sat down in the sunshine. The grass was soft and fragrant, filling the air with a feeling of peace and tranquility. Merlin rolled around, purring loudly with pleasure.

"This reminds me of home," Arthur remarked easily. "This lake feels just as magical as the one in Avalon."

"Avalon," Marissa breathed. "I wasn't sure if that was real or not."

He looked at her quizzically. "Real? Of course, Avalon is real. It's where Merlin lives."

At the mention of his name, the little cat mewled softly.

Arthur laughed, and said, "No, not you, little one." Then he continued. "Yes, Avalon is a grand and wonderful place. I've only been there once, but Merlin speaks of it often. It holds many secrets, secrets I hope to discover sometime soon. Merlin tells me it is not yet time for me to learn everything, but I grow impatient."

"I can understand that," Marissa agreed. "You are king and need to know everything about your kingdom."

"Those are the exact words I have said to Merlin, yet he puts me off. It is aggravating. He tells me my destiny is to rule Britannia, yet he holds back information. If I did not love him so, I would surely think he is plotting something against me."

Marissa turned to Arthur and vehemently stated, "Don't think that. Merlin's only thoughts are of you. He would never do anything to put you or your kingdom in any jeopardy. He is loyal to you."

"I am sure you are correct," he said. "Yet, it is still frustrating, him holding secrets."

"I think everyone around us thinks they know better than we do," Marissa remarked. Arthur nodded in agreement. They sat in easy silence for several moments, allowing the tranquil feeling of the lake and the autumn day to relax and soothe. The sound of Merlin's purring was accompanied by a bird's song wafting across the air. A splash could be heard out in the water, and a bullfrog began to croak along the bank.

Sitting quietly, Marissa's attention was drawn to a soft humming sound. She looked around, and saw a huge, green dragonfly buzzing lazily next to her. The transparent wings were a rich jade color, and its long, graceful body looked like it was made of emerald. She gently raised her hand, and the dragonfly easily alit on her outstretched finger. It was the largest dragonfly she had ever seen, measuring at least three inches in length. She looked at the beautiful little creature, totally awed by the fact that it would perch so willingly on her hand.

Before she realized it, another dragonfly appeared next to her. This one was a bright red, with garnet wings and a ruby colored body. It too was huge as dragonflies go, and it also perched upon her hand, sitting next to the emerald.

Arthur whistled softly beside her, and said, "This is truly a magical place."

She smiled at him, nodding. "I've never had this happen before," she said. "They aren't afraid at all."

While she studied the two beautiful insects, yet another dragonfly appeared. This one was bright orange. Then another, sapphire blue, buzzed close. The two sitting on Marissa's hand took flight, and in moments, it seemed thousands of dragonflies were flying, swooping and darting all about the two monarchs. None flew as close as the first two, but they all seemed as if they were putting on a show for the royal audience of two. Every color of the rainbow filled the air around Arthur and Marissa, as dragonfly after dragonfly flitted this way and that. The sun glinted off bodies of opal, topaz, amethyst and every gemstone one could imagine.

Even Merlin sat up to watch the spectacle, as thousands of dragonflies filled the air with dazzling light. Marissa was reminded of a fireworks display as she watched the bright colors dance through the air, but this was so much more exciting than mere fireworks. This was nature at its finest, bringing to life the colors one saw all too infrequently in the everyday, mundane experiences of daily existence.

Dragonflies flew together in a cluster, looking like a huge, multifaceted jewel. They then dispersed and spread out, bringing color this way and that, filling the air with sparks and flashes of gemstone light; green here, blue there, red and orange darting together. The beauty was breathtaking, and Marissa and Arthur watched in silent awe as the creatures seemed to be dancing just for them.

The show went on for several minutes, then suddenly, the dragonflies drew into a tight ball of swirling color. With a pulsing motion, the ball of color exploded, scattering the dragonflies in every direction, and the dragonflies disappeared. Neither Marissa nor Arthur saw where they went, they were simply gone. Merlin let out a melancholy yowl, and lay back down.

"I agree," Marissa said to the little cat. "That was beautiful, and over much too quickly. I am truly amazed each and every day at the magical occurrences of this land."

"Yes," Arthur agreed. "I grew up working with my hands, cleaning stalls, bathing horses for my knight. Since I drew Excalibur from the stone, my life has changed completely. Every day something happens that makes me realize how very little I know, and I appreciate so much more the wonders of the world."

"I know exactly what you mean," she said. "I know I can rule my kingdom, and I will protect my people. But the world is full of so many things I don't understand. I try to keep an open mind, and see everything with a child's eye. When I stopped questioning everything, I grew to treasure every new experience."

Arthur nodded. "I too know I will rule well. It is my destiny, and I will not fail. But I still have a lot to learn." He looked at Marissa, a question in his eye. "Could that be why Merlin gives me information and knowledge in small pieces?"

"So you're more able to absorb it and learn it, instead of being overwhelmed?"

"Yes," he said, nodding. "Perhaps I finally understand why he takes his time in teaching me." He chuckled. "I have been annoyed with him on many occasions when he refuses to tell me everything. But now I see. Perhaps I can slow down somewhat, and persuade him to speed up somewhat."

"That's a good compromise," Marissa said. "You have time to learn everything."

"Perhaps," he said darkly. "And perhaps not. The lesser kings have been at war for generations, and I want to put a stop to it and bring them all together. That is my mission, my destiny."

"And you will succeed," she assured him.

"But I do not have all that much time. With each new sunrise, more men die needlessly. The wars should be stopped quickly. I need allegiance from the kings, from the kingdoms."

"Well, you have mine," Marissa stated firmly. "I am your servant, and my kingdom supports you completely."

"That is good to hear," he replied. "With you pledging your support, hopefully the other kingdoms will follow suit. It has not been easy. I expected them to understand that I only want what is best for Britannia, yet they seem suspicious of me. It is as if they believe I have an ulterior motive, that peace and unity are not enough motive to put a stop to the fighting."

"They will see," she said. "You will bring peace to the land, and they will all follow you. Just give them time to see that you will deliver what you promise."

Arthur laughed softly. "It seems you know more of me than I do myself. Your faith in me is most welcome."

"Ah, Arthur. You have no idea of the great things that will come your way." But to herself, she thought, "unfortunately, it won't last."

TWENTY-EIGHT

There was yet another celebration to say farewell to the new High King. Marissa walked through the crowds, silently musing about the fact that there seemed to be more banquets than any other activity in Avonridge. Well, when something is done well, there's no reason not to do it often.

Sir Erick walked beside her, and she turned to him and said, "I am glad King Arthur and I made an agreement. I feel much safer and more protected knowing the High King regards me as an ally."

"It is good to have the favor of the High King," the black knight agreed, and then was silent once again. Marissa still wished he would converse more, she found it frustrating not to have someone to discuss the everyday matters of the kingdom with. But, alas, she reminded herself that he knew his place, as medieval as that was. And considering they were living in the medieval world, she must accept the trappings of royalty and rank, no matter her private feelings.

As they walked, the ladies-in-waiting came rushing up to Marissa. She groaned inwardly, realizing she'd completely forgotten to have them assigned to other duties. With rescuing Arvin from Balvindor, and readying the palace for King Arthur's arrival, she had, quite understandably, put the ladies out of her mind. Now they clustered around her, clucking and pecking like so many barnyard hens.

"Majesty," the youngest one, Louisa, gushed, "please tell these other brainless wenches that I will be sitting beside the High King at the evening meal."

"Majesty," another lady, Jane, cut in, "please tell this young twit that she is no more fit to sit beside the High King than is a sow from the sty."

"Majesty," yet another stated, "please tell…"

"Enough!" Marissa shouted, and the ladies all gasped, and became silent. She glanced at Sir Erick, and could swear he was smiling behind his leather mask. Turning to the ladies, she stated, "I can't believe what I am hearing. You are all fighting over who will sit beside the High King? Do you think any of you will?"

"Well, of course, Majesty," the youngest said. "He is in need of a wife. Who better than a lady-in-waiting from your court?"

Marissa looked at the young woman. She was only about as old as Arthur himself, and she could understand the girl's eagerness to gain favor in his eyes. After all, being the wife of the High King could be an extremely comfortable job. Unfortunately for Louisa, Marissa knew exactly whom Arthur would eventually marry. In fact, that wedding should be taking place in the very near future, if she had her facts straight.

Shaking her head, Marissa tried to regain her composure, and ease the ladies' anxieties about the banquet.

"Ladies," she said gently, "it is impossible for any of you to sit beside the High King." All the women groaned in disappointment, but Marissa continued. "As you know, I will be sitting on one side, and Merlin will be sitting on the other. You all will have the opportunity to come up and speak to him. In fact, I will gladly encourage him to spend some time with each of you. Only a moment or two, but at least you will be able to have some time with him. Is that sufficient?"

"Well," Louisa groused, "if that is the best you can do, I suppose it will suffice. I did so want to sit with him, though."

"I'm sure you did," Marissa said soothingly. "Unfortunately, protocol must be observed. I myself grow tired of all the things I must endure," and she heard Sir Erick give the slightest chuckle, "yet I try to live up to my subjects' expectations. Now run along and prepare for the evening's festivities." And she quickly stepped past the group of clucking women and hurried away before they could annoy her any further.

Taking long strides to widen the distance as quickly as possible, she glanced at Sir Erick. "You fully enjoyed that," she stated snidely.

"I am quite certain I do not know what you mean, Majesty," he said easily, and she could hear the laughter behind the words. She laughed herself, shaking her head at the mundane annoyances she had to tolerate. She definitely had to speak to Ruth about getting rid of the flock of women that made her blood pressure rise so drastically.

Walking through the crowd of revelers, Marissa wished King Arthur would stay a while longer. However, she understood his need to continue visiting the other lesser kings. Now that he had a strong alliance with Avonridge, he hoped the other kings would follow suit. Bringing peace to Britannia was a priority, and the sooner he could achieve it, the better for all involved. Marissa herself hoped it would happen before any other king decided to attack Avonridge. After all, Sorens couldn't be the only one who coveted the fertile lands and prosperous kingdom of Marissa's reign. Every minute wasted was another chance for an attack.

"There you are, Majesty," a familiar voice called. Marissa turned to see Forsythe striding up to her. "I was hoping to spend some time with you."

"Oh, really," she replied, only a hint of sarcasm in her voice. "I would have thought you'd be trying to bend the High King's ear, not mine."

"Ah, of course," he said easily. "I do wish to speak with His Majesty. Unfortunately, I realize I am but a minor player in the game of life, and must wait until he can spare a few precious moments of his time."

She looked at him, and once again remarked, "Forsythe, you really should be a poet. Your command of the language is quite impressive." But she was actually thinking that his manner was fawning and unctuous, and saw through it like a clear glass window. He smiled, and bowed slightly at the compliment.

Sir Erick was hanging back just a bit to give Marissa and Forsythe some privacy in their conversation, and she glanced back just to make sure he didn't stray too far. She had come to enjoy Forsythe's company, at least when the alternative was a group of infuriating women whose only concerns were for their own greedy designs.

However, as much as she welcomed the conversation from a man that treated her as his peer, that very fact made her uneasy. She didn't trust this man that raised his own social status, or worse yet, lowered hers, so to consider them equals. But it would be good to take advantage of the respite from the tedium of royalty.

As they walked, Forsythe clasped his hands behind his back in a comfortable posture. "It has been a while since we've had the opportunity to chat," he said. "I have missed being in your company."

Marissa didn't answer, just continued walking.

"I fully enjoyed having Arthur here, he's brought some stature to our humble little community. Do you agree?" His voice made it clear he fully expected her to agree.

"Actually," she stated easily, "I don't feel our kingdom is humble. I am quite impressed and pleased with its stature."

"Well, of course I mean no disrespect," Forsythe said, laughter softening his words. "I consider myself quite lucky to be a subject of Avonridge. She is a prosperous and beautiful kingdom. I simply meant that Arthur brings his own prestige. With his presence, we gain stature in the eyes of the other kings."

"Of course," she replied, patronizingly. "However, I find it curious that you would address the High King as 'Arthur'. I was under the impression that you were familiar with palace protocol, and would never overstep your boundaries." Her voice remained soft and melodic, but the meaning was crystal clear.

"Yes, again I mean no disrespect. I suppose I feel comfortable with the High King, due mostly to his youth. He seems quite the easygoing young boy."

"Whatever he seems, he is still the High King."

"Of course." They walked in silence for several steps.

"I am quite pleased with the efforts of all who cleaned the castle and courtyard," she said, changing the subject. "I am proud of the presentation we made for the High King. I hope he sees the care and diligence that was put into getting the kingdom in order for his visit. I would be quite dismayed if he didn't realize the respect we have shown him."

"I would not put too much store in that, Majesty," Forsythe stated. "He is but a boy, and boys do not pay too much attention to the cleanliness of castle walls."

"I suppose you're right," she agreed. "But it would have been disgraceful if he found the walls dingy upon his arrival."

"Quite disgraceful, I am sure," he said.

They continued chatting and strolling for a while. Marissa made sure the conversation was about nothing, and found her mood lightening from the encounter with the ladies-in-waiting. Presently, Forsythe excused himself, and strode away. As soon as he was out of earshot, Sir Erick resumed his place beside the queen, and said, "His arrogance is insulting."

"Oh, I can tolerate it," Marissa replied. "I know exactly the kind of person he is, and he's enjoyable in small doses."

"But he did not even bow upon encountering you, nor upon leaving you. If it were mine to decide, I would run him through at the very least."

She was silent, realizing for the first time that he was exactly correct. Forsythe had not shown respect to her, his familiarity was becoming too familiar. She needed to remember her place as queen equally as much as her subjects needed to remember their places. She mentally reprimanded herself, and decided it would not happen again.

As she and Sir Erick walked, a cold knot formed in Marissa's stomach. She recognized that feeling, and looked around to find Marcellus. Off to the side, in the shadow of a stone wall, she spied the stooped figure of the man that hated her. She stopped and gazed in his direction, simply staring. The figure remained motionless, but she could feel the hatred and animosity flowing from his concealed eyes. She glanced at Sir Erick, who seemed oblivious to Marcellus' presence. This still chilled her, but she took comfort in knowing the old man probably couldn't sneak up on her.

She strode toward the malevolent figure, Sir Erick keeping pace with a quizzical look in his eyes. As she approached the alleyway where he stood, she could feel the knot in her stomach churn and roil. She stopped a few feet from the old man, and Sir Erick sucked in his breath as he spied him. Marissa said, "If you have business with me, Marcellus, have at it. I do not care for your skulking about."

"I do not skulk about, Majesty," came the icy reply. "I simply do not enjoy crowds the way you seem to."

"I see," she challenged. "But you do wish to speak to me?"

"Perhaps," he said. "Perhaps you should be made aware of circumstances that would harm the sanctity of your kingdom," and the final two words were spat out.

"And what would those circumstances be, old man?"

He paused, looking off into the sky over her head. After a moment, he spoke. "I am enjoying watching your reign slowly fall apart," he sneered.

A chill ran up Marissa's spine, but she refused to allow him to see her reaction. "I can't possibly know what you are talking about, old man."

He chuckled, a raspy winded sound. "You trust the wrong people, Majesty," was all he said, then turned and hobbled away.

"That man never ceases in his attempt to undermine your reign," Sir Erick stated uneasily.

Marissa nodded agreement. "Yes, but there's something more going on with him," she said. The Black Knight turned to her with questions in his eyes. "He knows something," she continued, "and he actually just made me aware of a potential problem."

"He simply wishes to cause you to doubt everyone in your orbit," Sir Erick insisted. "He did not make mention of just whom you should distrust, he said only enough to cause you to worry."

"But I won't worry, I will simply be more cautious. I will keep my eyes open and not let anyone surprise me."

Sir Erick nodded, but said nothing more, obviously unsatisfied with Marissa's lack of concern.

TWENTY-NINE

As the evening wore on, Marissa spotted King Arthur milling through the crowds. A pang of sympathy stabbed her heart as she also saw three of her ladies-in-waiting clustered around the king. She could see all three talking at once, leaning toward him as they walked, each trying to catch and hold his attention. For his part, Arthur was smiling and nodding, obviously trying to be polite, but with a cornered-animal look in his eyes.

Marissa watched for a few moments, then decided that, as good as King Arthur was at dealing with cantankerous old kings, he had very little experience in dealing with romance-hungry young women. She decided to rescue the poor boy.

Striding up to Arthur and the three ladies, Marissa called, "Your Majesty, I was hoping I could have your attention for a few moments."

He turned to her, relief so stark on his face she had to suppress a smile. "Of course, Majesty," he nearly shouted. "How may I be of service?"

The ladies-in-waiting still gathered around, unwilling to relinquish any of the king's attention, even to their queen. She realized subtlety would do no good, but still didn't want to hurt their feelings.

"Ladies," she said sweetly, "I have important matters to discuss with the High King. I'm sure you would find it all excruciatingly boring. Why don't you enjoy the banquet for a while?"

All three grumbled and complained quietly, but did in fact leave. When they were out of earshot, Arthur sighed and said, "Thank you. I wondered if I would get out of that alive."

"Yes, they can be very trying," Marissa agreed, smiling.

"I am glad you found me for another reason as well," he said as they walked. "I spent some time talking with a man who claims to be your 'suitor'."

She glanced sideways at him, and he was looking at her quizzically. "That's how he identified himself, as my suitor?" she said.

He nodded slowly. "That is how."

"Let me guess," she said dryly. "It was Forsythe."

"It was."

Sir Erick, walking a step behind, scoffed. Marissa looked at him, and knew nothing needed to be said. She fully understood the Black Knight's feelings toward Forsythe. However, arrogant no longer seemed adequate to describe this man.

"So," she said to Arthur, "in claiming to be my suitor, exactly what did he ask of you?"

"At first I thought it was all quite innocent. He spoke of you with the lilt of a lover's voice. I found it quite refreshing, really. He continually complimented you on your victory over King Sorens. He remarked on how you've battled the dragon more than once, and been victorious each time. He really spoke as if he loves you."

"But," Marissa said, "you don't completely buy it."

"Buy it?" he asked, confused. Then seemed to understand. "Oh, yes. You mean 'believe it'."

Marissa groaned inwardly. She still had some things to get used to here in this magical land.

"No," Arthur said, "I in fact do not buy it. As I said, at first it was lovely to hear. But then it began to sound contrived. And soon he began to extol his own virtues. In fact, he spoke of himself unceasingly for several minutes. Until finally, he asked me to persuade you to marry him."

Marissa stopped and sucked in her breath. "He didn't!" she said.

"He did," Arthur replied.

She stood still for several moments, so caught up in incredulous anger she couldn't speak.

Arthur went on. "I assured him I could not involve myself in intimate matters such as this. I would never infringe upon personal decisions like marriage."

"And how did he respond to that?" she asked, finally finding her tongue.

"He said it was not a completely personal matter, as the marriages of kings and queens greatly affect the working of the kingdom."

"That arrogant son of a …" she trailed off. Then she began to laugh. "I do not believe it," she said, still laughing.

"You find it humorous?" he asked. "I don't understand."

"Oh, I am just amazed that someone can have such nerve," she said. "It's a shame he is not a soldier, he would make a good one with that kind of audacity."

"Do you think so?"

"Perhaps," she said. "But then again, with his total disregard for possible consequences, he'd probably get himself and his whole army killed."

"That is more my feeling," Arthur agreed. "His brashness comes from total and complete self-service. He seems not to have the best interests of anyone but himself in mind. That kind of man can cause a great deal of havoc."

Marissa looked at this young man she had become so fond of, and smiled. He seemed so naïve, so young. But his insight was admirable. He was no pushover, and he would make a great king.

"So, did he try to convince you my kingdom would be better off with a king, instead of a queen?" she asked.

"I believe that was his intention, however he never actually stated it. At that time, I feel he realized how delicate the subject had become."

"Yes, he doesn't want to be accused of treason by doubting the capabilities of his queen," she said. "After all, his ambitions are quite high-reaching. It wouldn't do him good to make me, or you for that matter, angry with him and suspicious of his intentions. And I expect he'll tread lightly now for a while. I am sure he is hoping you've reconsidered and decided to try to persuade me." She laughed again.

"You have a good attitude about it," he said. "I dare say I would be angry were the situation reversed. If someone sought my king's influence in my choice of a bride, I would not tolerate it well at all."

"Well, I want to see the humor in it now, because if I don't, I will get angry. And right now, I want to enjoy the evening. The air is cool and comfortable, the smells of roasted meats are enticing, and the company is most enjoyable," she said, smiling at him. "I don't want to spoil it. There is plenty of time to deal with Forsythe later."

And they continued strolling through the night, changing the conversation to more pleasant topics.

The army of Avonridge was lined up along the walls leading out of the castle gate. Flanking the road King Arthur would be taking as he continued on his quest for the allegiance of the kings, sat knights dressed in gleaming armor astride horses decked in colorful drapes of red and gold. The banner of Avonridge, with the depiction of the dragon in battle, flew majestically on both sides of the road. Outside the open drawbridge, the soldiers of Avonridge stood in formation on either side of the byway leading away from Avonridge.

Marissa stood upon the drawbridge as Arthur and his army approached. She was dressed in gleaming white, and sitting majestically upon her head was the golden crown with the carving of the dragon's head. Her purple cape framed her petite form, and she looked exquisitely regal as she prepared to bid God-speed and farewell to the High King.

Arthur sat astride his mount and slowly drew up to the queen. His own armor reflected the sun in blinding glory, and clanked softly as he moved. His captain of the guard and Merlin the Magician each rode beside, and a pace behind, the king, with Arthur's own standard-bearer proudly carrying the flag of the Pendragon.

Arthur stopped in front of Marissa and dismounted. Walking to her with an easy, confident gait, he stopped and bowed deeply at the waist, as the queen curtsied to the ground.

"Queen Marissa," he stated in a strong, clear voice, "I thank you for your hospitality. I have felt most welcome here within your kingdom."

"And I thank you, King Arthur," she said, rising, "for the honor of your presence. It has been a great joy to be host to you and your magnificent army. I truly hope we can meet again and continue our friendship."

"As do I," he said. Then he bowed again and remounted his horse. Marissa stepped to the side as King Arthur led his army across the drawbridge and down the road between Marissa's soldiers. The soldiers of Avonridge all held their right forearm across their chest in salute to the High King as he rode out of sight.

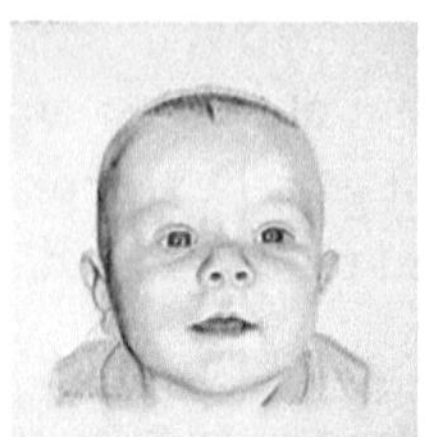

THIRTY

Again, life seemed dull at Avonridge. It seemed there was little to do between celebrations and battles. Marissa thought it odd that she was not called upon to do more mundane activities around the castle, but knew Ruth was taking care of the everyday running of things.

Weeks had passed since her victory over King Sorens, and she was still housing the imprisoned soldiers of his army. She called Roland to her chambers to discuss the terms of returning them to their home. Knowing she could not simply release them and allow them to go, she wanted Roland to give her points on proper bargaining. He was happy to help her, and quite adept at the process.

What she should do, he'd said, is to send word that Marissa was willing to negotiate for the return of his soldiers. She would make all the demands, offering no choices. She had approximately two thousand men of Southcross, and that should be a large enough bargaining chip. She would let him know her terms, what she expected in exchange for the soldiers. Roland offered the example of wheat or corn, something that would prove useful for the coming winter.

When Sorens was given the ultimatum, he would undoubtedly send back a counter offer. Considering the arrogance of the king, he would probably first insist the soldiers be returned with no recompense. However, he would most definitely then offer something of little or no value. Marissa would do well to ask for an unreasonable amount, allowing room for negotiation. It should all be quite easy, a simple process. The men of Southcross should be home within a month or so.

Marissa was pleased with Roland's suggestions, and a soldier from Marissa's army was dispatched to Southcross. It would take a few days to reach Sorens, make the offer and wait for the reply, so she went about the everyday life of the kingdom.

Approximately a week after King Arthur had departed, she was finally called upon to perform a duty that was not dangerous, or boring. She would sit in judgment for all sorts of disagreements her subjects faced. She was the only court in Avonridge, and when there were matters to be decided, they were brought before the queen on the first night of every full moon. Marissa found this interesting and challenging, as she knew her decisions would be binding, and possibly life-changing to some people. She felt the responsibility, and strove to be fair, impartial and reasonable in her judgments.

On the first night of this full moon, the harvest moon, she was hoping for a gradual entry into the legalities of disputes. She mentally prepared to hear quarrels over ownership of livestock, or other such petty troubles. Unfortunately, she would not be allowed to learn to swim in the kiddy pool, she was about to be thrown into the middle of the lake.

The first dispute was over the death of a woman in labor. A grieving widower was calling for the execution of the midwife he believed responsible for allowing his wife and son to die in childbirth. The man, a youth really, was distraught beyond consolation, and sought to punish the woman he felt caused the deaths. He accused the midwife of neglect when his wife's labor continued for four days without bringing forth the child. He claimed the midwife should have been able to facilitate the birth, and therefore spare both mother and child.

The midwife, an old crone of a woman, was also distraught over the deaths, genuinely affected by the tragedy. Unfortunately, as the midwife pointed out, women and children still die in childbirth. It is a dangerous, and even deadly, endeavor. The baby had not turned, and so was breach. There was little to do but wait and pray. With each contraction, the midwife helped the mother push, hoping to speed the process along. Nothing seemed to help, however. She claimed to have done her best, but God, not she, was the ultimate decider.

Marissa thought hard about how to handle this delicate situation. Finally finding and idea, she called for other women who had used the services of this midwife. There were a few in the audience, and each seemed eager to be a part of the drama, rushing forward to tell their tale. For the most part, each woman was satisfied with the midwife's performance. However, for every three or four happy mothers, there was one who'd lost a child, or given birth to an imperfect baby.

Marissa knew these odds were not unreasonable in this kind of society. With practically no sanitation, absolutely no technological amenities, labor and childbirth would be a voracious killer of women and children. As heartbreaking as that was, it was reality. Wanting desperately to ease the man's pain, there was really no justifiable reason to honor his request of executing the midwife. So, with a heavy heart at the man's devastating loss, she found relief in not passing final judgment on one of her subjects for her first trial.

The harvest moon had waned, and the air grew ever cooler. The trees along the hillside turned magnificent shades of crimson, gold and umber. The sunrises through the crisp morning air were breathtaking, lighting the sky in scarlet and orange. Marissa grew excited over the coming holiday season. She was still not completely up to speed with all the traditions of this magical kingdom, but she was enjoying learning as she went along.

As September ended and October began, the soldier she'd sent to Southcross returned. He seemed no worse for wear, and quite a bit confused. The king had refused to see him, completely ignoring the messenger from Avonridge. He would not grant audience, so Marissa's offer could not be presented. After two days, word had been given to the soldier to return to Avonridge, or face execution. Of course, the choice had been an easy one.

Marissa could not figure this one out. Why would a king not bother to try to reclaim his own soldiers? She had seen his cold-heartedness during the battle, knew of his arrogance, but would he really turn his back on his soldiers? This was not the action of a fair and just king, this was the action of a despot.

Well, she would figure out what to do about it in due time, right now, in the late fall she wondered about Halloween. Would this primitive land celebrate something that unordinary? But then, she remembered that Halloween had begun as a pagan holiday, and even though Avonridge was most definitely a Christian community, there were still very vivid and solid references to the old pagan ways. Would Halloween be the Beltane festival?

No, that was in May and would coincide with Easter. What was the name of the festival held in the fall? Oh, she would find out soon, she was sure. Yes, Halloween might be called something else, but it would certainly be celebrated in one way or another. And Marissa was extremely excited at the prospect of seeing the children of the kingdom dress up and dance for the celebration.

Marissa loved Halloween. Even as an adult she enjoyed the dressing up and becoming a different person, even a different entity. Of course, that was how she'd gotten involved in the Renaissance Faires in the first place. But Halloween was practiced and accepted throughout the whole country of the United States. The Renaissance Faires were still looked down upon by a great number of people, and so she'd kept that part of her life secret from her coworkers. Funny, it had only been a few months, and yet that life seemed like a distant and almost forgotten dream.

She met with Ruth, and began to make preparations for the celebration. Samhain, that was the name of the autumn festival. She knew something about it, but needed to learn everything she could.

Samhain, pronounced sow-in, is celebrated on the first of November. This marks the end of the harvest season, and the beginning of the cold, dark winter, the time of year most associated with death. The death of both flora and fauna, it also included human death. Winters are harsh and cruel in a world where the people rely on hearth fires to keep them from freezing to death.

Within this celebration, it is believed the ghosts of the dead return to the earth to bring knowledge of the future for Druid priests to decipher. This knowledge would help the Druid priests make prophesies and predictions for the coming months of harsh cold and inadequacies. On the night of October thirty-first, the Druid priests would dance around an immense bonfire, singing to and calling the spirits of the dead, asking for their insight and guidance to sustain the people throughout the winter months.

Marissa found this ritual quite dark and dreary, and overpowering for the children of the kingdom. She would allow the traditions of Avonridge to continue, as she believed they still had their place and function, but she would also begin a new tradition, the Halloween Party. Planned to coincide with the bonfire rituals, a huge party would be held in the palace for all the children of the kingdom, along with their parents.

Ruth reminded Marissa that the palace could not possibly hold all the children of the kingdom and their parents. Common sense winning out, Marissa realized Ruth was correct, and decided to hold the party within the fortress walls, throughout the whole of Avonridge proper.

Marissa sent riders to every village within Avonridge to announce the coming celebration, and ensure every subject was invited. Preparations were begun and Ruth directed the household staff in organizing exactly what would be needed for the celebration. Pies, cakes and cookies would all be baked; candies and confections would be created; and of course, meats would be roasted. Again, Marissa thought about all the feasts Avonridge had hosted since her arrival.

As the month of October marched forward, wood was brought and erected throughout Avonridge proper in teepee style for the bonfires the Druid priests would need. Huge tables were placed along the streets of Avonridge to hold all the treats her subjects would be offered. Marissa walked through the streets, enjoying the feeling of the air as it grew steadily colder. She had always loved the cold weather; heat and humidity sapped her of enthusiasm and energy. And so, as she walked, she breathed deeply of the cool, fragrant air, feeling invigorated and excited about the future.

King Sorens' soldiers were again put to work making the kingdom ready for the celebration. Roland and his soldiers kept a watchful eye on the prisoners, but none seemed unhappy or restless in their toil, again seeming to enjoy the exercise and fresh air. Marissa made sure they had been well taken care of, having decent food and comfortable accommodations, even if they were housed in the stables, and now they seemed happy enough to perform menial tasks in exchange for the continued civil treatment.

She could only guess at the treatment Sorens offered his people. With a cold shiver, she remembered the cruel way in which he had used his soldiers as stepping-stones to escape the suffocating marsh mud during their battle. As king, he certainly needed to save himself, yet the way in which he did so was, to Marissa, most foul.

Continuing along the streets, surveying the progress being made in readiness for the celebration, Jasper, the young page hurried up to her. She remembered him as the boy that had brought her to meet Queen Gertrude the day she arrived in Avonridge. His homely face and lanky clumsiness had not been eased by the months of growth he'd experienced, he still reminded her of a mule.

"Majesty," Jasper called, as he dropped into another clumsy bow, again almost toppling over, "I have been asked to bring news to you."

"News," Marissa echoed. "What news?"

"The man you rescued from the dragon, what was his name," this last more to himself, "Arbor, Albert…"

"Arvin," Marissa provided.

"Yes, Majesty," he gushed, relieved, "Arvin. He has asked me to bring news."

"Yes, Jasper," she said, amused at the boy's nervousness, "what news does Arvin ask you to bring?"

"Oh, ah, yes," he stammered. "Oh, now I remember! He said you told him no child would ever be unwanted."

Marissa waited for him to continue, but Jasper seemed to think he had said all he needed to say. "And, is that the only news he asked you to bring?" she prompted.

Jasper looked lost for a moment, then realization became evident in his homely face. "Oh, yes, he said to give you the news that a child has lost his parents."

Marissa's heart gave a lurch of pain. "Oh, how tragic. But Arvin remembered my promise. Jasper, where is Arvin now?"

"He waits at the drawbridge, Majesty, with the little boy."

"Bring them to me at once," she commanded, and he bowed quickly, then spun on his heel and practically ran away. So, there was a child in need. She paced slowly back and forth, waiting for Jasper to return with Arvin and the boy. How terrible for a child to lose both his parents. She didn't even know what age the child might be. Would he be old enough to understand, and therefore be scarred by the loss?

She continued pacing, thinking how to go about providing a family for this child. As queen, she could order a family to care for him, but could not order that family to love him. That would be a great disservice to both the child and the family. She knew of no orphanage within her kingdom. But, of course, that was a modern institution, having come about primarily during the industrial revolution, which would be many centuries in her future.

No, she needed to figure out how to help this child now. She thought of Arvin's son Darvin, who was seven years old, and supposed if this boy was old enough to be able to help with farming and other duties, it would be easier to find someone to take him in. After all, he would be able to help pay his own keep.

But that still wouldn't ensure the boy would be loved. Having grown up with parents who loved her dearly, Marissa knew full well the value of feeling wanted and needed. She must do her best to provide that basic necessity to this poor child. After all, the subjects of Avonridge were her children, and she promised to care for each and every one.

As she paced, she finally spied Jasper bringing Arvin and the boy. She had been thinking of Darvin, and expected the orphan to be about that age. What she saw, however, surprised her. Arvin was carrying a toddler, no more than two years old. As they drew near, Jasper excitedly leading the way—it seemed that wherever he went, he went excitedly— she watched the tenderness with which Arvin held the child. Several thoughts raced through her mind as the three made their way toward her; perhaps Arvin could be persuaded to care for the baby, perhaps a childless couple within the kingdom would welcome him, and finally, the boy was adorable!

Jasper led Arvin to Marissa, bowed, and stepped back. Arvin also bowed, reverence and awe in his clear blue eyes. "Majesty," he said softly in greeting.

"Arvin, it is good to see you again," she said, smiling. "I'm delighted to see you so well after your ordeal."

He smiled self-consciously, and said, "I am no worse for wear, Majesty. Thanks to you I have been able to continue tending my cattle and raising my son."

"And how is Darvin doing?" she asked warmly.

His chest swelled slightly and his smile became loving as he said, "He prospers, Majesty. He is learning quickly how to tend the cattle on his own. He grows stronger and smarter by the day." The obvious pride in his son was a joy to witness.

"Wonderful," she exclaimed. "It must be a great comfort to know he will be able to manage, no matter what life throws at him."

"Yes, Majesty," he readily agreed.

"So, now," Marissa said, "who is this little cherub?"

The boy in Arvin's arms had been studying Marissa from the moment he saw her, a tentative smile on his round little face. His eyes were deep brown, with green rings around the irises, making them sparkle. His short-cropped brown hair was filthy and plastered to his head, but waves could clearly be seen beneath the dirt. His face was equally as dirty, as were the rough cotton overalls he wore. Despite the cool weather, he wore no shirt or shoes, the sleeveless overalls being his only attire.

Unlike the denim overalls Marissa associated with American farm workers, these were rough, brown cotton, resembling burlap, and cut from one piece of cloth sown along the sides and legs, almost as if a burlap sack had indeed been fashioned for his clothing. He seemed to sense the conversation had turned to him, and he gave a grin that outshone the very sun itself. Marissa's heart melted.

"This is Kevin, Majesty," Arvin announced, "son of Evin." And he smiled at the boy.

Marissa stared at the tyke, feelings crashing through her like the cars of a demolition derby: pity at his loss, concern for his welfare, and sympathy for his uncomfortable and insufficient attire. "He's a beautiful child, Arvin," she said softly.

"That he is, Majesty," he agreed. "But he has no family."

"Tell me what happened," she asked.

Arvin looked at Kevin, bouncing him affectionately. He obviously cared about this lost little waif. "His kin lived at the edge of our village," he began. "Evin and Jonelle had three daughters before Kevin came to them." He hesitated, seeming to gather his thoughts. Marissa waited patiently.

"Just two nights ago," he continued, obviously distressed, "while the family slept, a fire escaped from the cooking hearth. It was hungry, burning everything it could touch. We could hear Jonelle crying out for help. We rushed to their shack, trying to help, but the fire was angry, it lashed out the door at anyone who came near." Arvin took a breath and lowered his head. Marissa could sense the anguish he must be feeling, knowing they couldn't help the family.

In a hollow voice, he resumed his narration. "We fought the fire with blankets and buckets of water, but it was too strong, too vicious. We could not subdue it before it destroyed the home. We heard the screams of the family as they were eaten alive by the fire. That night will haunt me for many years, Majesty." His voice had become a sob, and a tear rolled down his dusty cheek.

"But Arvin," she said gently, "if they were all killed, how did Kevin survive?"

He looked up at her, his eyes filled with wretched pain, and her heart broke anew. "A mother's love," he whispered, his voice filled with awe at the memory. "As we tried to bring down the fire, we saw something struggling to get through the door. It was a ball of flame, walking by its own power. It reached the door and collapsed, burning until we doused it with water. That was when we saw it was a blanket wrapped around a person. That person was Jonelle, with Kevin tucked to her breast. Even though the blanket burned, and Jonelle burned, we were able to kill the fire before Kevin was burned. Jonelle shielded him with her own body to save him."

They stood in silence for several moments, the only sounds being Jasper's soft sobs at the tragedy. Marissa had almost forgotten about the page, but he stood off to the side venting the feelings she and Arvin struggled to contain.

After a few minutes, she asked, "Doesn't anyone in the village want to take him in? Isn't there a couple who can treat him as their own?"

Arvin took a moment to answer. Then, speaking slowly, as if to choose his words carefully, he said very quietly, "The child is cursed."

Marissa wasn't sure she'd heard correctly, and said, "What was that?"

He looked at her, pain still clouding his eyes, and repeated, "The child is cursed."

She shook her head. "I don't understand. What do you mean, 'cursed'?"

Dread colored his face a ruddy red, and he shrugged miserably, but said nothing.

"Arvin, why is this innocent child cursed? He's done nothing to bring the wrath of anyone down on him."

"He survived," Arvin said simply.

"Again, I don't understand," she insisted.

He seemed not to want to explain, but she continued to stare at him, waiting. Finally, he said, "Majesty, you do not understand, because you are protected here at the castle. Curses cannot touch you with your army and your magic. But we…, we are not protected. We have no magic to fight the curses, we must only try to hide from them, stay away from them."

"Arvin, you aren't making sense. Tell me why anyone would curse this child."

"We do not know why, but the fire was sent by some evil force. If anyone takes this child into their home, they will then have the curse upon their family. Fire goes where it is sent, and it was sent to destroy Evin and his family. We do not know why, we can only try to stay away from it."

Marissa stared at the simple peasant man, so wracked by fear of the curse that sent the fire. But she could also see affection for the child in his eyes. He wanted the child to be cared for and protected, yet knew no one in the village would dare take him into their home. He obviously believed the curse couldn't reach Kevin here at the castle, he'd said as much by mentioning her army and her magic. Wouldn't he be disappointed to know that she herself possessed no magic?

She shook her head again and sighed. "Arvin, don't worry about Kevin," she said. "I will see to it that he's cared for and loved." She reached out her arms and the boy eagerly moved into her grasp. Arvin's face relaxed into a smile of relief and gratitude when he saw how comfortable Kevin was in her arms. The boy was smiling at Marissa and softly patting her cheek. She swayed with him, looking every bit the doting… what? Aunt? Mother?

Marissa thought about how she'd never had children of her own. And Ruth told her the queens of Avonridge did not bear children, the risk was too great. Having presided over the dispute with the young widower against the midwife, Marissa fully understood the danger in this society.

She had, without realizing it, resigned herself to the fact that she would never bear her own children. But a child had been brought to her in need of a family, a child who had suffered greatly at the hands of cruel fate. She was never a true believer in coincidence, she believed things happened for a reason. Could Kevin have been brought to her for a purpose?

She knew the kingdom would not be handed down to a male successor. The traditions of Avonridge were there for a reason. But Kevin most likely wouldn't live long enough to ascend the throne anyway. Since he hadn't been fashioned from magic like the queens of Avonridge, he would live a normal lifespan. So, the problem of not passing the crown to him would be moot.

Could she actually be considering this? Could she actually adopt this little waif? If no one else in the village, probably even the whole kingdom, would dare bring a 'cursed' child into their home, what other option did she have? She looked into the sparkling brown eyes that were surrounded by such a dirty face, and she knew. From this moment forward, Kevin would be her son.

THIRTY-ONE

Preparations continued for the Halloween celebration. As Marissa walked through the streets of Avonridge, Kevin trotted happily at her side. Tiny fingers were wrapped tightly around the first two fingers of her right hand, and she walked slowly to accommodate his toddling steps. She had tried to carry the tot, but he would have none of it, insisting on his own ambulation. He allowed her to hold him and cuddle him, but when they were going anywhere, he would plant both little hands against her shoulder and push with all his might.

He didn't speak, not a word, but Marissa wasn't too concerned about that. At two years old, he should be chattering away, but considering the circumstances by which he had come to live with her, she wouldn't push him until he was ready. She kept up a steady conversation with him, however, hoping to coax out of him some verbal response.

Just as she had suspected, when bathed and clothed in decent attire, Kevin was a most beautiful child. His curly hair framed his round face, and those deep brown eyes simply danced with delight at every new sight and experience.

Having been born into a peasant family and spent his entire two years in a tiny shack in a tiny village, he had never seen all the different things the palace had to offer, and he delighted in every new color, every new sound, every new activity. Being a bold and adventurous child, he eagerly reached for shiny objects like the silver cups and gleaming white porcelain plates used to feed him. He smiled almost continuously, and waved 'hello' and 'good-bye' whenever he was instructed. But he still would not utter a word.

Walking through the streets of Avonridge, Marissa beamed with love and pride at the beautiful little cherub walking beside her. However, she became annoyed, and even a little angry, when she remembered the reaction from the ladies-in-waiting when she brought Kevin to her chambers. Being of the gentility of the kingdom, they were quite snobbish, and did not want even to touch the toddler.

Of course, he was still filthy dirty when she'd brought him in, calling for the bathtub to be brought and filled immediately. But, even considering his physical condition, he was still an innocent child needing to be cared for. The ladies-in-waiting backed away from him as if he'd displayed all the physical symptoms of the plague.

Marissa instantly berated them for their arrogance and apathy toward a child that was not to blame for his circumstances. Unfortunately, they continued to shy away from him, refusing to allow him near them, and looking at him with ill-masked disdain.

Merlin, however, had been a shining light for Marissa in his acceptance of Kevin. Immediately, the little black cat strutted up to the toddler, wrapping his feline body around the boy's legs and meowing in greeting. Kevin laughed and grabbed a handful of Merlin's coat, tugging painfully, but Merlin only purred, completely allowing the boy his fun. Marissa could almost swear she'd seen a look of pained resignation on the cat's face, and laughed out loud.

Unfortunately, Ruth had had almost the same reaction as the ladies-in-waiting when she'd first seen Kevin. She clucked her tongue at Marissa, quite disapprovingly, but had said nothing. When Kevin was washed and dressed, Ruth had smiled at the little waif, and even though Marissa knew it was forced, had accepted him into the palace household.

Walking through the streets, Marissa remembered the conversation she'd had with Ruth when she told Ruth that Kevin would be staying.

"It certainly is your decision, Majesty," Ruth said, keeping her voice neutral. "Whatever you think best."

But Marissa heard the strain in her voice, as Ruth knew she couldn't dissuade her from keeping the boy. Advising the queen on royal protocol, and all the different nuances of palace life, Ruth knew the way things should be, and also knew the way things were. And Marissa, though an adept pupil in learning the royal duties and expectations, still had a mind of her own. And so, Ruth acquiesced, saying no more on the subject.

As Marissa walked through the streets of Avonridge with Kevin at her side, subjects greeted her all along the way. Some asked about the tot, and others simply looked at him quizzically. For those who asked, she stated easily that he had come to stay with her, and didn't elaborate. She didn't need to hear the disapproving clucks she knew would come her way when the people found out the details.

For the most part, even if questioning her about Kevin, the people of Avonridge were more interested in the coming Halloween feast. Commenting on the smooth assembly of the tables and bonfire piles, they expressed excitement at the idea of another celebration, so conversation was easy and superficial.

Walking along, Marissa noticed Kevin tugging on her hand. She looked down at him and saw a look of anxiety, even fear, on his little face. She looked in the direction that he was staring, and noticed Forsythe making his way toward her. As he drew nearer, Kevin seemed more anxious, trying to pull Marissa in the opposite direction, away from the tall man in the brightly colored green tunic, and smug expression.

"I know exactly how you feel," she muttered softly to the little boy. "But don't worry, he can't really hurt either one of us."

"Majesty, it is good to see you," Forsythe greeted unctuously as he approached. Marissa noticed again that he did not bow. Well, there would be time to address his lack of protocol soon enough. She didn't really want to go so far as to call it a lack of respect, however, she knew in her heart that was exactly what it was.

"Forsythe," she said. "To what do I owe the pleasure?" she asked, only a hint of sarcasm coloring her voice.

"Oh, the pleasure is mine, Majesty," he quickly replied. "I am glad to have run into you. It always lifts my spirits just to be in your presence."

She turned her head in the pretense of looking at the preparations, but really to hide the look of annoyance she felt. She had also noticed that he hadn't paid the least bit of attention to Kevin, not even asking who he was.

"So, Majesty," Forsythe continued, "this is going to be a wonderful celebration. I hear there will be all sorts of sweets and treats for the young ones."

"Yes," she agreed, "we are planning on spoiling the children. And speaking of children, have you met Kevin?" she asked, indicating the little boy who was doing his best to hide behind Marissa's skirt.

Forsythe's eyes glassed over and his face adopted a wooden expression as he looked at Kevin. "Well, I had heard you'd brought a child into the palace. It is a shame you didn't discuss with me that move before you'd made it."

Marissa raised her eyebrows at him. "Excuse me?" she said, somewhat surprised.

"Yes, Majesty, I could have counseled you on the wisdom of that decision."

"And exactly what council would you have given?" she asked, her voice growing hard with warning.

Forsythe, however, either didn't hear, or chose to ignore it. "Well, as you know, the queen of Avonridge must keep up appearances. Bringing a child of questionable parentage into your household…, well, I must say, it does not look particularly good to the neighboring kingdoms. Do you not agree?"

"No, I do not agree!" she stated. "I have no interest at all in what the neighboring kingdoms think about me or my realm. And as for the 'questionable parentage', he is a child of Avonridge. That makes him thoroughly acceptable to anyone within Avonridge. Do *you* not agree?" she demanded, the warning now unmistakable.

"Do not misunderstand me, Majesty," he replied, an arrogant lilt to his voice. "I have no such reservations. It is simply a matter of good judgment. I am simply questioning the wisdom of such a decision."

"Questioning?" she echoed, her voice now a razor's edge. "Forsythe, would you have me believe you would question any decision your queen makes?"

His face colored slightly, the arrogant lift of his chin lowering, as realization became evident on his face. "Ah, Majesty," he said, "please forgive my unfortunate choice of words. Of course, I would never question any decisions you might make. I am only trying to protect your good name and reputation."

"And exactly how would my reputation suffer by bringing this innocent child into my household?" she demanded.

"Ah," he stammered, shaking his head in denial of any intentional misstep. "Majesty, surely you understand my position. I assure you, I am only being vigilant for your welfare. As a loyal subject of Avonridge, it is my duty to protect you in all things, and we do have a relationship you do not share with everyone. After all, we are so close, I feel a special kinship to you."

Marissa stared, mouth slightly agape. Kevin was still tugging on her hand, trying to hide completely behind her skirt. She looked down at the little boy, then back at Forsythe. Finally, she found her tongue. "And to exactly what kind of kinship are you referring? I am sure I do not understand. I have never felt any closer to you than to any other of my subjects. Why would you feel there was a closeness between us?"

He let out a short, derisive laugh. "Ah, Majesty, there is no need to pretend, no one can hear us. There is no one around to take offense at the depth of our relationship. I am simply referring to the closeness I have felt to you since your arrival. Since the first moment I laid eyes upon you, I knew there would be a special bond between us. And this little boy, cute as he may be," he said, glancing down at Kevin with obvious disdain, "simply has no place within the palace household. He is a commoner, and should remain with his people, not those of us that are higher born."

Now it was Marissa's turn to laugh. "Really Forsythe, you amaze me each and every time we cross paths. Are you still under the impression that you might persuade me to marry you?"

"Direct and right to the point," he said. "I like that. In answer to your question, I have never given up hope that you would indeed consider my offer. As I said, since the moment I saw you, I knew I would love you forever. It is only a matter of time before you realize the benefits of our union."

"And exactly what would those benefits be, may I ask?"

He placed one hand upon his chest, fingers thumping softly, and the other one behind his back, taking a stance that made Marissa think of old Hollywood pirate movies. "Just think about it, Majesty," he said. "The running of the kingdom can be a heavy burden. With me by your side, you would not have to make the difficult decisions all by yourself, you would always have a companion to discuss any circumstances…"

"But I have that already," she stated. "I have my advisors and my friends. So, you see, I do not need you at all."

"With all due respect, Majesty, exactly what friends do you have? There is none within the kingdom that shares your rank and position. You are quite alone in that regard. But with a suitably born husband, you would no longer have to carry that burden alone. You would be able to share the responsibilities and the duties. Now, there are a few subjects with high credentials, but, if I do say so myself, none with quite my degree of parentage. I am sure you would find no finer choice in a husband."

Marissa was shaking her head, a small smile of disbelief upon her face. "Correct me if I am wrong, Forsythe," she said, "but didn't we already have this conversation? I seem to remember several occasions, actually, where I informed you not only would I not be taking a husband, but, even if I did, it would certainly not be you."

His smile faded, and was replaced by an ill-concealed mask of contempt and anger. He recovered quickly, however, and put his wooden smile back in place. He laughed lightly, as if she had just told a cute little joke, and said, "Of course, Majesty. I am again being too forward. I will simply bide my time until you come to realize just how advantageous it would be to accept my offer. I am sure you will do so quickly, as you are an extremely intelligent woman. So, until then, dear lady, I take my leave." And without a bow, he turned on his heel and strode away.

Marissa watched him go, confusion mixed with anger churning in her stomach. Was Forsythe really arrogant enough to believe she would consent to marry him? She had met some pompous men in her life, but none quite as pompous as this man.

Kevin gave a tug on her hand, bringing her attention back to him. She looked at his upturned face, a face that clearly showed fear. She picked him up and gave him a cuddle, and he wrapped his tiny arms around her neck and buried his face in her shoulder.

"What is it, little one?" she asked softly. "Does that man really frighten you?"

Kevin lifted his head and looked into her eyes, and she had her answer.

THIRTY-TWO

"Majesty, I must protest, I simply must protest!" the bishop nearly shouted. He was still dressed in the scarlet robes and miter hat of Marissa's coronation, apparently this was his usual dress. He ran up to Marissa as she walked with Kevin, and, panting, repeated, "I must protest."

"Bishop," she said easily, "please calm yourself. Tell me exactly what it is you protest."

Other than her coronation, Marissa had only seen the bishop in the cathedral on Sunday mornings as she attended mass, and had never seen him agitated as he so obviously was at this moment.

"Majesty," he stated in a pious and righteous voice, "Avonridge is a Christian kingdom. It does not hold with the old Pagan ways. You yourself attend mass every Sunday. You know Avonridge should not be perpetuating the Pagan ceremonies."

"Bishop, I am fully aware that we are a Christian people," she said calmly. "However, the Pagan ways are very much a part of our history, and to ignore them would be a great disservice. We can respect the old rites and still not compromise our Christian beliefs."

"I do not agree, Majesty," he argued. "Celebrating the old ways is a direct insult to our Lord and Savior, Jesus Christ. He has given up his life for our souls, and to repay him by celebrating Pagan rituals is an abomination. He would be very displeased, very displeased indeed. To allow the old Pagan ways to flourish anew is simply unacceptable. We cannot condone, encourage even, the old ways."

"Bishop," she stated firmly, "by celebrating Pagan rites, we are not giving our lives over to the Pagan ways. It is still a reality that some of my subjects are Pagans, and to allow them their right to practice their beliefs is only fair and just. We are not reverting to Paganism."

"Majesty, you are not seeing the whole situation," he persisted. "If we allow the Pagan followers to practice their beliefs, it is only a matter of time before the whole of Avonridge is corrupted and thrown back in time. We have progressed far since embracing Christianity, and to allow the old ways to regain power…"

"Bishop!" she cut him off. "We are not abandoning Christianity. We are simply celebrating one day. True, it began as a Pagan ritual, but do we, as Christians, not celebrate All Saints' Day?"

"Of course, we celebrate All Saints' Day, Majesty, that is not the issue here."

"But why not think of it as allowing the whole kingdom celebrate All Saints' Day? We are bringing all the people of the kingdom here for a celebration. We are not in reality celebrating the dead, we are celebrating their souls. We are remembering them."

"But Majesty," he protested, but his voice grew less righteous.

"Now, Bishop, if we are remembering the souls of our ancestors, the souls that reside in Heaven with our Lord and Savior, how can this celebration possibly be looked upon as anything but a loving and joyous occasion?"

He paused, clearly searching for a viable argument, but could find none. He shook his head in confusion and anger, and Marissa gently laid her hand on his arm.

"Bishop, we are not embracing the Pagan ways. We are simply remembering them as a very real part of the past that helped create Avonridge. There is no need to feel threatened by the celebration."

As they spoke, Druid priests were arriving and inspecting the wood for the bonfires. Marissa could see the bishop grow uneasy again at the sight of them, but she cupped his elbow in her hand and turned him away from the sight. If she could get him away from the location that was so agitating him, she may be able to get through the Halloween party without any further incidents. She decided to change the subject.

"Bishop," she said, "you are aware of how Kevin came to live in the palace, are you not?"

"Of course, Majesty. News of that sort travels quickly." And he leaned down and patted the little boy on the head, a genuine fondness obvious in his manner. This made Marissa's heart soar.

"So, since you know something of his history," she continued, "would you know if Evin and Jonelle ever had him baptized?"

"The poor of the kingdom believe very strongly in baptizing their children," he said. "I was not called to perform the rite, but I am quite certain a lesser priest would have been."

"But there is no record of the baptism?" she asked.

"We have no such records, Majesty," he said, as if that were the most obvious of statements.

"So, we can assume, but not be sure, that he was baptized."

"I suppose that is correct, Majesty," he agreed.

"Good. Then, just to be sure, I would like to plan his baptism with you, if you would be gracious enough to perform it."

The bishop looked at Marissa, a small smile upon his wrinkled old face. "That would be a most enjoyable duty to perform, Majesty," he said sincerely. "We have not had a royal baptism in the palace in quite some time. In fact, I believe the last was for the lady Louisa. Yes, I cannot remember any since hers. It was a grand celebration, as well. Her parents were so proud. And she was a beautiful little child. But she is quite the adult now, is she not?"

"Is she ever," Marissa groused quietly. She remembered the lady Louisa trying in vain to get the attention of King Arthur. Of course, she was no worse than any of the other ladies-in-waiting; arguing, pecking, sniping, all in their attempt to curry favor with the High King.

"And so, is the second weekend of November a good choice for the date?" she asked the bishop. "That would give us a couple of weeks to prepare."

"Yes," he answered. "That would be an acceptable time for the occasion. Winter would not have settled harshly upon the land by that time, making travel fairly easy for the guests coming from afar."

"Guests?" she echoed. Yes, she would have to invite people. Ruth would be able to advise her in the obligatory guest list. She hoped it wouldn't have to be too extensive, she was actually tiring of all the banquets she was hosting. It would be so nice to have just a small ceremony, with a little luncheon afterward. But, of course, as queen she must be ostentatious.

The evening of October thirty-one was cool, clear and brightly lit by a million stars. The bonfires were roaring at different spots throughout the streets of Avonridge, and the tables were laden with roasted meats, pitchers of wine and apple cider, and sweets of all imagining. Cakes stood between platters of cookies and bowls of chocolates.

Children ran through the streets, screeching with delight, laughing and grabbing cookies and candies off the tables as they ran. Marissa walked through the castle grounds in a bright orange gown of silk taffeta, with black ribbons adorning the sleeves and bodice. Having expected costumes of ghosts, witches and monsters, she was surprised to see only a handful of handmade bunny-like ears here and there.

Obviously, the Halloween she had grown up with was centuries away from this little kingdom. When instructed to spread the word of the Samhain festival, Sir Erick had returned expressing the questions and confusion about the costume idea. No one had ever had a costume party in Avonridge, and so no one knew exactly what to make of it. He had explained about dressing up as another person or animal, hence the bunny ears, but that was the extent of it. Besides, she knew peasants couldn't really afford to get their hands on costumes.

Oh well, Marissa was delighted anyway, the children of Avonridge were having a great time enjoying the party. The bishop, however, was anything but delighted. Marissa had expected him to remain in his rooms at the back of the castle, away from all the "Pagan" festivities, but he was bustling through the crowds, tsk tsking, with every step. His face was drawn and pinched as if he had a taste of sour lemons in his mouth, and he was shaking his head at all the laughing children as if he expected them to turn into horned devils and run away at any moment.

She almost felt sorry for him, he was so distressed with the unChristian-like behavior that his face was a ruddy purple and his breath came in short gasps. "Disgraceful, simply disgraceful," he muttered to himself as he walked past Marissa. "God-fearing Christians behaving like heathens, I do not know what this world is coming to." But the children were having such a good time, she decided to ignore the bishop's discomfiture.

Kevin was tugging on Marissa's hand as they walked through the streets, trying to get loose to go play with the other children. She let go of his hand and watched him dart toward a crowd of little ones playing a ring-around-the-rosy game. When they saw him approach, they stopped dancing in a circle and one little girl grasped Kevin's hand. The little boy on the other side of Kevin also grasped his hand, reforming the circle, and they resumed dancing, dragging Kevin along with them. His shrill laughter joined the other voices, and Marissa dabbed at a tear of happiness at the sight of him being accepted into the group.

With the party well underway, and two ladies-in-waiting assigned to watch Kevin, Marissa was free to wander among the revelers. She walked to where a Druid priest was performing his ritual dance, and was captivated by the fervor and passion he exhibited. The priest's long dark robes billowed out and seemed to dance in the air has he spun, clapping his hands and dipping his head in time to music that rose softly from a stringed instrument being played several feet off to the side.

The musician was sitting cross-legged, with what appeared to be a lute in his lap, head bent low in concentration. The music was eerie and haunting, barely audible above the din of the crowd, and it didn't seem to fit the exuberance with which the priest was dancing. But he dipped and twirled around the huge bonfire, chanting words that she couldn't understand.

She turned and caught sight of the bishop, still shaking his head and looking as if he was watching his whole congregation being thrown into the pits of Hell. She chuckled under her breath, and continued walking, watching similar bonfires and similar dances here and there along the streets. The bishop would just have to endure it.

Ruth had done a wonderful job of setting the mood for the evening. Great orange and black silk drapes hung against the castle walls and the walls of the shops along the streets. Marissa came upon a great barrel filled with water, and several apples floating within. Older children, ranging in age from about seven to eleven or twelve, were gathered around it cheering as one girl was bent over it, hands holding back her long hair, trying to catch a floating apple in her teeth. On the far side of the barrel, Marissa spotted the young boy Darvin, and when he spied her, he smiled and waved. She waved back and continued walking.

Fully enjoying the festivities, it was several moments before she realized there was a knot of ice in the pit of her stomach. She recognized the feeling immediately, and looked around for the source. Standing off to the side, between two buildings and almost completely hidden, she found the one she had been looking for.

Marcellus was deep in the shadows of an alleyway, unseen by everyone, everyone except Marissa. She could just discern his form, stooped and malevolent, but she could feel his eyes upon her. Not wanting to spoil the mood of the evening, she decided to ignore the old man, and continued on her way. Unfortunately, he had different ideas, and stepped from the shadows, walking directly toward her.

She took a deep breath, bracing herself for the verbal duel she was sure would begin any moment. He hobbled toward her, and she wondered again at his age. He seemed as old as the very earth upon which she stood, and yet still had an air of strength.

"Majesty," he sneered, and dropped a half-bow. "I see you have rekindled the old Pagan rites. It is good to have a sovereign who is willing to return to the tried-and-true ways of Avonridge."

"We are returning to nothing, Marcellus", she said easily. "This is just a celebration in honor of a part of Avonridge's past, nothing more."

"Ah, but the bishop seems quite distressed over nothing more than a celebration," he challenged. "If the Pagan ways are not being reinstated, why is he so distraught?"

Not wanting to get into another discussion of the Christian versus Pagan rites, she simply stated, "You would have to ask the bishop. I cannot answer for him." And she turned to continue her walk. Unfortunately, Marcellus was not content to let her go.

"I understand you have acquired a ward, Majesty," he said.

She looked into his dark eyes, not sure what she saw in them. "Yes," she said. "I have been given a little boy to care for. Why?"

He only chuckled, causing the hairs on the back of Marissa's neck to stand on end. He seemed to be toying with her, and her anger began to simmer.

"Don't tell me you want to protect my reputation, Marcellus," she said with open hostility, "and advise me against bringing a commoner into my household?"

"I would advise nothing of the sort, Majesty," he said, and chuckled again.

She opened her mouth to demand to be told what he found amusing, but at that moment Kevin rushed up to her, Lady Jane and Lady Louisa close behind. The two ladies-in-waiting had been none too happy about their assignment of watching the little boy, but they were at least taking their responsibility seriously, they stayed near him at all times.

As he ran up to Marissa, she scooped him up into her arms and cuddled him, allowing the closeness of him to comfort her. He turned his little round face toward Marcellus, and reached out a hand, smiling.

Marissa stared at the little hand, and then looked at Kevin to be sure he was in fact looking at the old man. He was, and to her further consternation, Marcellus was smiling back!

She studied the old man's face for any sign of malice or duplicity, but found none. The smile on the wrinkled old face seemed completely genuine. He did not, however, reach for Kevin's hand, to Marissa's great relief, but Kevin was clearly not afraid of him. Not only was he not afraid of Marcellus, he seemed to like the old man. Dumbstruck, Marissa simply looked back and forth from Kevin's smiling face to Marcellus' in wonder.

The absurdity of the situation gave Marissa a chill, and she turned her back on the old man and strode away, ladies Jane and Louisa hurrying to keep up.

"Ridiculous!" she scoffed at no one in particular. "This little angel is afraid of Forsythe, and yet seems to prefer Marcellus." She looked into Kevin's eyes, so big and filled with enjoyment of the evening, and asked, "Do you really like that man? Please tell me you don't really like him."

He only grinned at her.

THRITY-THREE

With Halloween over and Kevin's christening to look forward to, Marissa had not a moment's rest. Of course, Ruth arranged all the details, and drew up the list of whom to invite. However, the thought of yet another huge banquet was becoming a burden. Could this really be the way life at medieval court had been? Was there nothing to do but host banquets? Well, if the only alternative was to fight wars, banquets were certainly preferable.

As the day drew near, lords and ladies began to arrive at the castle. Rooms had been prepared in anticipation of guests staying for a week or two. It was good to be able to meet all the nobility of the kingdom, and even a few neighboring kingdoms sent representatives. Even though most of the kings were involved in the wars, some were friendly to Avonridge, and Marissa welcomed them. She even spoke to the queen of Caldwell who seemed to despise King Sorens and Southcross as much as she did herself.

"Yes, that man has been causing trouble for as long as I can remember," Queen Sadie mused as she and Marissa strolled through Avonridge. She was an older woman, somewhat plump, with soft gray hair pulled back in a chignon. Carrying herself with dignity, she was rather handsome, with the regal air of having been born into, and then married, royalty. Marissa was quite pleased with herself that she wasn't the least bit intimidated by her.

The weather was cool and clear, and the sun shone down on the walls of the building, causing the polished white stones to gleam like glass. Kevin toddled between the two women, and Queen Sadie glanced down at him from time to time, smiling like an indulgent aunt. "I often wonder," she continued, "if the wars would not have ceased long ago if it were not for that arrogant man and his high ambitions."

"He does seem to want to rule everything around him," Marissa agreed. "The day of my coronation he issued a challenge and tried to seize my kingdom."

"Oh yes," Queen Sadie said. "That news spread through the lands very quickly. It was most welcome indeed." Marissa looked sideways at her, and she quickly stated, "Oh, not that he had challenged you. No, what was so welcome was the news that you had trounced him soundly. Oh, we were so proud of you, and happy that he'd been put in his place. And the fact that a woman had done it," she said, lowering her voice, "made it all the better."

Marissa nodded and smiled.

"Yes, that man is quite full of himself," Queen Sadie went on. "We believe he wishes to be High King. We would not be surprised at all if that is his ultimate ambition. Poor, dear Prudence. She suffers so much at the arrogance of that man. She has tried to convince her husband to abandon the wars, as you have done, Marissa, dear, but he ignores her every time. It is such a shame, Queen Prudence does justify her namesake, she's a lovely woman."

"If Sorens has designs on the throne of the High King," Marissa said, "he'll be quite disappointed. King Arthur is High King, and it won't be long before he puts a stop to the wars and brings peace to the land."

"Yes," Queen Sadie said distractedly. "If he's capable."

"You don't believe King Arthur capable of stopping the wars?" Marissa asked.

"Oh, I'm certain he has every intention of doing so," she said. "It is only a matter of, can he do it? He is so young and untried. I wonder if any of the older kings will even grant him audience, much less obey his orders."

"I wouldn't be too concerned about the kings obeying," Marissa stated. "King Arthur has a way of getting one's attention."

"But he must be so arrogant, to be so young and yet be High King."

"Not at all," Marissa assured her. "In fact, he's quite unassuming."

"Oh, that's right," Queen Sadie said, "you've already entertained him. Tell me, won't you, did you really pledge your loyalty to him?"

"Completely," Marissa stated immediately. "I truly want peace between the kingdoms, and I believe King Arthur is the only one capable of accomplishing that. And if my allegiance will prod the other kings into offering him their loyalty, then all the better."

"Yes," Queen Sadie agreed. "I believe that will influence the other kings. In fact, my husband King Gregory, has already agreed to grant King Arthur audience, solely because you have given your allegiance."

"That is wonderful!" Marissa gushed. "Hopefully, more will follow."

"I believe, once the other kings see some of the hostilities fading, they will also be willing to accept peace," Queen Sadie said.

"So how long has King Gregory been in battle?" Marissa asked.

"It seems forever. I almost can't remember what he looks like. We send messages back and forth, but it is not the same. I almost wonder how we managed to have our two sons," she said with a wry chuckle. "So soon after our wedding, he was off to battle. His father had just been killed, and he needed to take his place." She heaved a melancholy sigh. "Such a waste. The wars have not accomplished anything except kill off our men. I don't understand their arrogance. Do they honestly believe they can continue their fighting and killing and still have their kingdoms prosper?"

"I have never understood men," Marissa agreed. "It is almost as if they think they can do whatever they want, and it will just magically turn out alright. I expect they figure their women will just take care of everything." She looked down at Kevin, and said, "Promise me you won't grow up to be that shortsighted."

He grinned up at her, but said nothing.

"I think you are absolutely correct," Queen Sadie stated. "They do act as if someone will just follow around behind them fixing everything."

Marissa chuckled. Apparently, men hadn't changed throughout history. They continued walking through the streets of Avonridge, enjoying the autumn air.

The celebration for Kevin's baptism was an elaborate luncheon. It was held inside the palace, in the great banquet hall where Marissa's arrival banquet had been. Again, the walls were draped in satin, but now only the colors of soft blue and white adorned the hall. At the head table, a chair was set for the guest of honor, and he sat primly in a pure white outfit, smiling at the guests. He still had not said a word, but Marissa did not concern herself with that little fact. He was a friendly, happy child, and she knew he would eventually come around.

To Kevin's right sat Marissa, and to his left sat Ruth. Sir Erick stood off to the side and behind the table, keeping his ever-vigilant eye upon the queen and her ward. Marissa had asked the Black Knight to sit with them, but he had declined, again citing palace protocol. On either side of Ruth and Marissa sat the ladies-in-waiting, chatting among themselves as if they were the only ones in the room. Marissa looked over at each in turn, and tried to let go of the annoyance she felt toward them. They had been behaving somewhat better lately, and were giving her less reason to want to reassign them.

In the past few weeks, the ladies-in-waiting had acted less and less like a bunch of barnyard hens, clucking less at Marissa. She wasn't sure if it was simply because they wanted to avoid Kevin as much as possible, or because they were becoming more comfortable with her as queen, but she didn't really care. They were less tiresome and annoying, and that was a bonus, for whatever reason.

The lords and ladies of the kingdom were seated throughout the great hall, talking amicably with those around them. Several queens from neighboring kingdoms had accepted the invitation, and were all seated up close to the head table. None of the kings could attend the christening, because of the never-ending wars, but their kingdoms were well represented. It was good to see such a fine turnout, it gave

Marissa hope that King Arthur would soon be successful in bringing all the lesser kingdoms together in peace.

She talked easily with the queens, and was heartened to hear that none of the kings believed the wars should continue. The only drawback was that none was willing to be the first to withdraw, being fearful that the other kings would overrun their kingdom. Should one of the kings take his army and simply go home, they were afraid of appearing weak, and therefore vulnerable. Marissa was in a unique position, having the only kingdom that had not been involved for generations. She was still flying under the radar, and King Sorens' attack didn't seem to cause much more than a ripple in her near-invisibility. Because the other kings were so involved in the wars, they didn't pay much attention at all to someone who was out of sight.

Queen Elsbeth of Memoria echoed Queen Sadie's feelings about King Sorens, and blamed him completely for the ongoing wars. She believed if Sorens would cease his aggressions and hostilities, the other kings could call a peaceful end to the conflicts, and return to their kingdoms, as so many wanted to do. Unfortunately, every time it seemed the wars were waning, the attacks growing fewer and fewer, Sorens would launch a new wave of assaults, bringing the other kings right back into the conflict. It seemed to be general consensus that something had to be done about Sorens, but since the kings were preoccupied with the wars, no one could afford to take the time to figure out exactly what to do.

The kings themselves were not exactly sure which king was on which side. Even though Sorens seemed to be the antagonist, none of the other kings trusted each other enough to make the first move in offering a council. None was sure the next was not in league with Sorens, and wouldn't alert him to any gathering. Should they all be caught together, the possibility of an ambush was too great to chance.

In fact, all the queens agreed that their husbands were not even aware that they were attending the christening of Queen Marissa's ward. They all felt there was no need to trouble the kings with details that they could not control anyway. The queens knew they would be safe since Marissa had already trounced Sorens, and they all believed it a good gesture toward bringing all the lesser kings together under King Arthur. Marissa marveled at the politics involved in running a kingdom. These ladies all were the decision makers in their husbands' absence, and knew how to play the game to the best advantage. Alliances between kingdoms were necessary to build a peaceful land, and since the kings were not able to get them underway, their queens were trying to do just that.

Also rewarding, and a little amazing, was the fact that the queens had no animosity toward each other. There was no hostility between them, and the atmosphere reminded Marissa of a college sorority. The queens all knew each other, and chatted and gossiped like old pals. She felt that the camaraderie would be quite beneficial in helping reestablish ties when the wars finally stopped.

She thought about when the wars finally did come to an end. She saw Camelot in her mind's eye, and wondered how long it would take to build it. From her readings, she thought King Arthur was newly crowned when he began to build his court and the round table, but was starting to get foggy on some of the details. Oh well, it didn't really matter. What mattered was that it would all eventually come to pass, and the wars would stop. At least, for a while.

The day wore on, and the partying continued into the night. At eight o'clock, however, the guest of honor had to retire for the evening, but that didn't hamper the festivities in the least. The luncheon turned into a dinner, complete with wine and a band. There was some dancing, but mostly just sitting around, drinking and talking. The comfortable feel of the company was most welcome, as Marissa had no one in her usual orbit to really let her hair down with. She had no peers in her own kingdom, and found it refreshing to talk with a group of women who were completely on an equal par with her. Their openness and sociability made her feel more relaxed and happier than she had in quite a while.

THIRTY-FOUR

The next several days passed quickly. Marissa had companions to spend time with that treated her as an equal. She knew the visiting queens, although always polite and regal, pretty much spoke their minds to her, not being afraid of her reaction to what they might say.

Soon, however, they all had to return to their own kingdoms. One by one, she bid farewell to the women who had become her friends. They all pledged to continue to nag their husbands to stop the foolish wars, and return to their palaces. If only they could convince them to put aside the needless hostilities and focus their attention on their people, the lands would prosper and the kingdoms could unite. It was an ambition they all would work toward, and sooner or later, God willing, they would succeed. If only the kings had the sense of their queens, but they all agreed, boys will be boys, and playing war seemed to be a staple that lasted throughout history.

Marissa walked through the streets of Avonridge, after wishing God speed to Queen Sadie, the last to leave, and was once again wondering what she would do without her newfound company. She didn't have to wonder for long, however, because her Captain of the Guard came rushing up to her.

"Majesty, it has happened again!" Roland shouted as he hurried up to her. "I had thought it was over, but it is not, it has happened again."

"Calm down, Roland," Marissa ordered. "I have no idea what you're talking about. What has happened again?"

"The Vulture's Claw, Majesty," he said, anxiously fingering the hilt of his broadsword. "The Vulture's Claw has again claimed a squad of my soldiers."

"What is the Vulture's Claw?" Marissa asked. Her anxiety was growing as she recognized true fear in Roland's demeanor. "And what do you mean it has claimed a squad?"

He forcibly calmed himself, and explained. "The Vulture's Claw, it is haunted. It takes men who dare to venture there."

"What is the Vulture's Claw, Roland?" she demanded.

"An evil part of the forest, Majesty," he said. "I sent a squad to patrol, to reconnoiter, in the event an enemy king may plan a surprise attack."

"That's good, we need constant vigilance," she agreed.

"But the Vulture's Claw is once again preying upon my men. I thought it had had its fill, but it seems to be hungry once again."

"I'm still not following you, Roland," she said. "What do you mean by 'hungry'?"

He took a steadying breath and continued. "The Vulture's Claw is a section of the forest on the eastern edge of Avonridge. It is deep and dark, the perfect venue to hide, or stealthily plan an attack. It had lain dormant for years, not taking anyone for quite some time. I mistakenly believed it was safe to endeavor there, but I made a gross error in judgment, and it has devoured my squad of soldiers," he finished miserably.

"What do you mean 'devoured'? Do we need to send a rescue party to save the soldiers?"

"It will do no good, Majesty," he said. "It is too late for them. It is all my fault," he whispered, hanging his head in despair.

"Roland, tell me everything," she insisted.

The Vulture's Claw, a large expanse of the forest, was so named many generations ago, because the only living things to brave it were vultures. No one ever witnessed another creature within its depths; not deer, foxes, nor even birds. Everything living gave it a wide berth. Even the vegetation seemed different; dark and evil. The trees were stooped and bent, rather than tall and majestic like the forest to the west. They grew leaves, but of a darker color, almost brown. Some claimed that they could feel a coldness about it if they strayed too close. No one was exactly sure why this was so, and some tried to explore it, discover its secrets.

The first squad of thirty soldiers was dispatched into the forest for a preliminary run through. They were to go in about a mile, draw maps, and return. After several hours, it was obvious they were lost. Being soldiers, no one was greatly concerned for their safety and knew they would eventually find their way back.

After two days, however, it became apparent the squad would not be returning any time soon. The people of Avonridge began to worry. A second squad was dispatched to search for the members of the first squad. The second squad was larger than the first, consisting of fifty men. They entered the forest at first light of the third day following the disappearance, eager to find and bring home their comrades. By nightfall, none had returned.

Debates ran through the army as to the wisdom of a second search party. No one wanted to abandon the missing men, but would it be foolish to risk more soldiers? Since there was no way to know what had happened to the men, they had no idea how to prepare.

Finally, concern for the lost men won out, and a troop of one hundred soldiers nervously began a new search. Townspeople stood side by side with the remainder of the army at the edge of the forest, watching as the soldiers entered. Walking twenty-five abreast, the soldiers would enter in four waves, each successive group keeping the previous one in sight. This seemed to be working well, until the fourth wave disappeared from view. All the spectators crept closer to the forest, while trying to keep out of reach of whatever malevolent force might be waiting to snatch them up.

After several minutes, sounds could be heard from deep within the trees, terrifying sounds, sounds of men screaming in terror. The Captain of the Guard quickly moved his men into formation and was ready to rush to the rescue, when silence descended upon the forest, and an icy wind blew out from within the darkness. Just as he was about to order the advance, the rustling of leaves and underbrush met his ears, and a lone soldier came staggering out. His clothes were ripped and torn, hanging in shreds, his weapons gone, and his eyes wild. He made his way to the Captain of the Guard, who reached to catch the soldier as he fell. The man lay trembling in the captain's arms.

"Tell me what happened," the captain ordered. The soldier said nothing, but continued to tremble, taking ragged, gasping breaths. "Soldier," the captain said, "tell me what happened. That is an order." Still the terrified man said nothing. The captain gently set him on the ground in a sitting position, and the soldier drew his knees to his chest, wrapped his arms around them and began to rock himself.

The captain shouted to his army, "Prepare to enter!"

At the command, the soldier on the ground finally responded. "No!" he screeched, fear causing his voice to crack. "No Captain, do not!"

The captain turned to him and knelt. "Solice," he said gently. "What happened in there?"

The soldier's haunted eyes locked on those of his Captain. "Evil," he whispered, then again became silent.

The captain looked into the forest. Devotion to his men would not allow him to ignore their danger. "Solice, we must try to help those men." He stood up and turned to lead his army, when Solice grabbed his arm.

"No, Captain," he pleaded. "'Tis too late to save them. They are gone."

"You know this?" the captain demanded.

Solice hung his head. "'Tis true," he said miserably.

"Why?" he asked. "Tell me what happened."

But Solice said no more. The captain was still inclined to enter the forest in search of the lost men, but the townspeople began to murmur.

"'Tis madness," one said. "Already dead," said another. "Certain suicide," called a third.

The soldiers all stood ready, willing to obey orders, but obviously terrified.

"You will be killing the entire army of Avonridge," called a man from the crowd.

"I cannot leave my men to face this alone," the captain answered.

"But you will be leaving the entire kingdom defenseless," came the answer. "We grieve for the lost soldiers, but we'll have no army if you lose the lives of all the rest."

The captain heard the wisdom in those words, and with an aching heart, withdrew his army from the edge of the Vulture's Claw. Solice had been taken home to his family, but never spoke another word. During the warm Avonridge summers, his wife would set him in a chair outside their little house, and he would stare off in the direction of the Vulture's Claw, with haunted, anguished eyes. His sons, only eight and ten years old when the tragedy had struck, tried in vain to reach him. They would sit beside him and talk about the daily activities of the village, or ask him questions about any number of things, but he never even seemed to know they were there. He was as lost as the men who hadn't returned from the forest.

For years, no one ventured near the Vulture's Claw again. Then some children decided to find out why everyone whispered when that part of the forest was mentioned. It had been over a generation since the disappearance of the soldiers. Most of the spectators had passed away. The only ones to have actually witnessed it were children at the time, and now were old and gray. The stories seemed all too fantastic to be true, the stuff of legends and folklore. Obviously, the old ones were losing their senses, and expounding on what was probably just a group of soldiers who lost their way.

Out of curiosity, and with the courage of youth, two young boys about ten years old cautiously crept to within yards of the trees. They strained to see into the gloom, but saw nothing unusual. They stepped closer, each prodding the other to be the first to enter. Finally, they summoned their courage, and stepped into the Vulture's claw.

A man spotted them as they disappeared within the trees, and ran after them. By the time he reached the forest, the boys were completely out of sight. He called to them, but received no answer. He stepped closer and called again. Finally, he heard laughter, and the two boys came running out.

The man was amazed. He ran back to Castle Avonridge proclaiming the Vulture's Claw safe. This brought a wave of curious spectators to see for themselves the forbidding forest. Scores of people lined up alongside the edge of the forest, peering into the depths of the trees. Cautiously, a few of the very brave stepped into the darkness, and hurried back out again. They emerged intact, and so others ventured in.

When no one came to any harm, and it became clear there was no danger, bands of hunters began to form, and deer and other game were sought. However, there was still no fauna to be found, and the flora made those who ventured there quite uneasy. When the townspeople realized there was absolutely no advantage to utilizing the Vulture's Claw, all activity finally ceased. It once again became a forgotten area, avoided now through apathy, rather than fear.

Several generations passed once more, and again curiosity grew. Since no one remained alive who had actually experienced the goings on of the Vulture's Claw, soldiers were once again dispatched, and once again they disappeared. As in the past, a rescue party was sent, and they too failed to return. History seemed to be repeating itself, as the second search party was lost as well, save for one unfortunate soul who had witnessed the destruction of his fellow soldiers, but was unable to give any details other than the presence of "Evil."

And so it went for the Vulture's claw. The stories were passed down through generations, until someone decided to prove the stories wrong. A period of evil and violence was followed by a period of peace, and then it would swing back again.

"And no one has ever discovered what caused the disappearances of the soldiers?" Marissa asked. "Not in all these years and all these incidents?"

"No, Majesty," Roland answered. "The only ones to live through it have been unable to relate what occurred."

"But what caused the violence to stop? Did it simply go away?"

"That I do not know," he said.

Marissa pondered what she had been told. It didn't make sense. Why would evil and violence occur, and then just fade away? Did someone do something to cause it to stop? Roland seemed to have no answers for the "why's", only for the "what". She looked at her distraught Captain of the Guard, and pain filled her heart, pain for her lost soldiers and their stricken families.

"There is no hope?" she asked gently, praying for the answer she knew she would not receive.

"None, Majesty," he said, his voice filled with anguish.

She remembered reading the Histories of Avonridge, and nowhere within the tome had she found mention of the Vulture's Claw. Did the previous queens not know if it? Had it been such a taboo that the subject was never formally recognized? These were all questions she needed answered, and she knew of only one person who might have those answers.

She bit back a sob, blinked away a tear, and said, "We will make provisions for the soldiers' families. They will not suffer any more because of this. I will set up a committee to keep an eye on them, and when they are in need, I will provide."

Roland nodded, as if expecting nothing less from his queen. She turned to go, and heard him whisper softly, "Gertrude has chosen well."

THIRTY-FIVE

"Camille," Marissa called, "I request an audience with you." The tiny hut had animal skins thrown atop the woven twigs and sticks, to keep out the cold of the approaching winter. The fire crackled and danced beside the hut, with a cauldron of bubbling liquid hanging over the flames.

She was patient, knowing if Camille was in there, she had heard the call. But Camille did not keep the queen waiting, and almost immediately the fur covering the doorway was pushed aside and the frail looking old crone of a woman emerged. It had only been a few months since Marissa had first met the old seer, and she had not changed in the least.

The colorful rags that covered her frail frame were themselves covered with furs. Winter had not fully arrived, but the air was cool enough to warrant an outer layer, and the old woman looked like a rich dowager with no sense of style. Mismatched furs hung about her shoulders and were tied around her waist; furs of fox, rabbit and something that looked amazingly like sable.

A soft smile warmed the old woman's face when she spied Marissa. "Majesty," she called, "please come and sit." She indicated the bench off to the side, and Marissa walked over and sat down. Merlin jumped up, stood in her lap and meowed loudly at Camille as the seer sat down.

"I have been expecting you, Majesty," she said, patting the cat's head. "You are here about the Vulture's Claw."

Marissa nodded. "I suppose I shouldn't be surprised. After all, you have the sight."

"Yes, and you need answers. This is a great mystery to you."

"It is," Marissa admitted. "I saw nothing of it in the Histories. This incident is the first I'm hearing of it, and I need to find out what happened, and how to put a stop to it."

"And only you can stop it," Camille stated. Merlin stepped onto the old woman's lap, and she gave him a quick hug. "And why have you not come to see me lately?" she asked the little ball of fluff. He meowed again, and she said, "Ah, I see. Much too busy for an old friend, huh?"

He looked duly chastised, and rubbed his head apologetically against her shoulder. She chuckled, and patted his head.

"Camille," Marissa began. "I don't understand what's happening in the Vulture's Claw. Why are there alternating periods of violence and quiet? What happens to make it start and stop?"

"Well, Majesty, what makes it stop is, the queen vanquishes the evil."

"Vanquishes the evil?" she echoed. "How?"

"Through magic," the seer answered. They sat in silence for several moments; Marissa staring at Camille, confused, and Camille serenely stroking Merlin's glossy fur.

Marissa shook her head as if trying to clear away cobwebs, and said, "Camille, please tell me what is happening in the Vulture's Claw, and how to combat it."

The old woman turned her warm gray eyes to the queen and said, "Of course, Majesty. The Vulture's Claw wasn't always as it is now, an evil place. Many generations ago it was just as serene as the rest of the forests surrounding Avonridge.

But one day people began to notice the trees were not as green as they had been. The leaves were turning darker, and the branches seemed stunted and diseased. Where once they had reached straight and tall, they now were becoming bent and gnarled. It seemed to be a process affecting the trees themselves, as it was happening to the ones already grown, and no new trees seemed to be emerging from the soil.

"The soil itself also seemed different. It had been rich and black, quite fertile, but was growing lighter, browner and more barren-looking. It appeared to have lost all its abundance, and looked as if it could not sustain any flora. The game that had once been plentiful now was absent. No longer could the townspeople hunt that portion of the forest to feed their families. The very air around it took on an evil quality, growing colder, even in summer.

"After a while the people of Avonridge demanded to know why the forest had changed, and called upon the army to investigate. The story of the disappearance of the soldiers has been passed down through the ages, but no real details were ever made known to the people. They took the knowledge of the events of the forest, and became too fearful to pursue the matter.

"When the soldiers' disappearance reached the ears of the queen, she knew she must put a stop to it. She came to me, just as you have done, and I gave her the knowledge and power she needed to vanquish the evil."

"Power?" Marissa said. "You mentioned magic. Did you give the queen magical powers?"

"A magical amulet," Camille replied. "Evil of that magnitude must be defeated by great magic, and only the queen herself can wield that magic."

She placed Merlin on the seat beside her, rose from the stone bench and ambled to her hut. She disappeared inside for only a moment, then emerged holding a brown wooden staff in her left hand, and something concealed in her right.

The staff was about five and a half feet in length, and an inch and a half in diameter. The bottom was flat and worn looking, as if it had been used as a walking stick. The top was smooth and rounded, as if it had been polished.

Camille stood before Marissa, and in a solemn tone, commanded, "Rise Queen Marissa, to receive the powers to save your people."

Marissa obediently stood, slightly confused and more than a little nervous. The anxiety faded, however, when she could feel a surge of power engulf her. The staff didn't appear to be causing it, so she looked at Camille's right hand. A white glow was emanating from between her fingers, and as she raised her hand, rays of light shone in every direction.

Camille placed her hand over the rounded end of the staff, and opened her fingers to reveal what appeared to be a large crystal marble. It was about three inches in diameter. She held it above the staff, and closed her eyes. Turning her head upward to face the sky, she intoned, "Powers of Avonridge, you are needed."

A tingle ran up Marissa's spine as Camille chanted and swayed. As the crystal glowed, she could feel the sensation of power increase.

"Powers of Avonridge," Camille continued, "the Queen's love for her people must be satisfied. Powers of Avonridge, you must awaken and serve the Queen."

The crystal shone brighter, a white blinding light. Camille slowly lowered it to touch the end of the staff, and the wood responded. The edge of the staff silently broke apart into three pieces resembling fingers. They curled up and around, latching onto the crystal like the prongs of a ring holding its gemstone.

Without a sound, the wood contracted, molding itself to grasp and support the crystal, fusing it to the end of the staff. When the process was complete and the crystal was secure, the brown staff itself began to change. It lightened in color, first looking like blond oak, then birch. When finished, it was a luminescent white, appearing almost ethereal.

Camille opened her eyes. She looked at the staff, then at Marissa. A smile spread across her wrinkled old face, and she leaned the staff forward, offering it to the Queen. Marissa grasped it and marveled at the coolness of the wood. With the light from the crystal and the whitening of the staff, she had expected to feel heat, but, even though it didn't look as such, it felt like ordinary, everyday wood. Except for the fact that she could feel a power within, like the force of electricity surging through it, and ultimately into her own body.

She twirled the staff around with her fingers, watching the rays of light glance off the trees around her. "I was about to ask you what I do with this," she said. "But it is already instructing me."

"Yes, Majesty," Camille replied. "All you need is right here," and she tapped the wood of the staff. "It will tell you everything you need to know. It is an old evil you will be facing, and this is an old magic with which to battle it."

Marissa nodded. "I can feel the spirits of the late Queens of Avonridge. They will help me."

"Each one has stood where you stand now," Camille explained. "Each Queen had to vanquish the evil anew. It seems to know when the old Queen has died, and takes advantage. It is one of the challenges for the new Queen."

Marissa understood, and she was ready.

THIRTY-SIX

Standing at the edge of the Vulture's Claw, Marissa peered into the gloom. It was indeed a menacing looking forest. The trees actually looked angry, as if resenting the evil that resided within. She could feel the coldness; it penetrated her skin to the bone and tried to grasp her heart. She pushed it aside easily, but knew that once she was within the depths of the Vulture's Claw, the evil would be stronger.

She carried no sword; her only weapon was the staff with the crystal. She had wondered if she would need her armor, but decided not to wear it. The support of the spirit of Avonridge was with her always. She felt surrounded, cocoon-like, by the strength and wisdom of all the queens that had come before.

"Well," she said aloud, "let's get this show on the road," and she stepped into the trees. It seemed as if someone had turned off the lights. Darkness enveloped her, and she blinked several times to help her eyes adjust. She quickly recovered, and looked around. It was dark and gloomy, but her sight wasn't at all impaired.

Immediately upon entering the forest, she felt a stirring. An angry wave washed over her, making her catch her breath. She could feel a presence, as if something had awakened, and she knew the evil was preparing to meet her.

She walked farther into the forest, anxious to be done with this business, and feeling quite uneasy. She would not give in to fear, however, as she was not alone. Even though they were not physically here, she knew the Queens of Avonridge were with her, giving her strength, and that knowledge warmed her and kept at bay the cold of the Vulture's Claw.

She thought of Roland. He was so very distressed when he learned of her intentions. He had paced the length of her chamber, arguing against the wisdom of her endeavor. The Vulture's Claw would swallow her up, he said, just as it had the soldiers. She must not go through with this foolishness, she was not expendable. Avonridge would falter, possibly collapse, should anything happen to her. Sir Erick echoed Roland's feelings, offering to go in her place if she truly insisted on this.

Ruth was just as distressed as Roland. She too tried to persuade Marissa to send someone else, or even let it go. The Vulture's claw was a dark and forbidding place, and no one need go in there. The townspeople would simply have to stay away. She and Roland refused to bend to her insistence of the necessity of her mission, their only thoughts were concern for her safety. She had left them standing at the drawbridge to Castle Avonridge, consumed by dread and fear, Roland pacing, and Ruth wringing her hands in frustration.

The forest seemed to swallow her whole as she continued deeper into its gloom. The presence of evil became almost physical, squeezing her heart, making breathing a struggle. She felt a sudden stab of fear that nearly buckled her knees, but she looked deep within herself, and recognized it as the first strike by her enemy. She forced the fear away, banishing it from her heart, and successfully turned aside the attack.

She knew she would not win this battle by defense alone, so launched her own first strike. Standing in the dark and dank of the Vulture's Claw, she grasped the staff tightly, took a deep breath, and announced aloud, "I will defeat you!" The crystal atop the staff sprang to life and glowed brightly, giving a warmth and light to the dark forest.

Marissa could feel, rather than hear, a reaction of pain from her enemy. Knowing this was not enough to achieve a victory, she resumed walking, with the crystal lending its light and power. A chill breeze caught her hair and billowed her skirt. Vaguely she realized this was the first movement of air she had felt, and it too had an evil quality. The whole forest resonated the evil of the being she was here to battle.

The crystal continued to gleam as if trying to ward off the evil it would face, the first line of defense. It didn't actually make navigating the path any easier, but gave Marissa a feeling of power, of strength, as she pushed forward. The breeze had ceased, but the chill remained.

The path was straight and narrow, as if purposely leading to a predetermined destination, and Marissa realized it most likely was. The only activity within the Vulture's Claw was the destruction of life, and it stood to reason the path would facilitate that. She walked without hesitation, feeling the draw of the battle that loomed ahead. This was another of her challenges, and she would meet it with the same determination and steadfastness with which she'd met the others.

It seemed like she had walked for miles, straight into the heart of the Vulture's Claw, if it had a heart. The forest was silent, devoid of the normal sounds of life--the scurrying of animals, the falling of leaves-- and the lack of any sound became disorienting. The only sounds were her own footfalls, which seemed hushed, as if the forest was trying to swallow them up.

After a while, Marissa realized the crystal was growing brighter. It glowed with a white opalescence. She felt the staff grow warm beneath her fingers, and knew she was nearing her destination.

The forest too had begun to change. The trees, still stunted and gnarled, were showing signs of some great force. Whether wind or a malevolent being, she could not tell, but here and there trees were uprooted. One in particular was somewhat awe-inspiring. It lay parallel to the path on which she walked, and its root structure was still attached.

The roots of the downed tree were still embedded within the earth in which it had stood, or rather, the earth was embedded in the roots. A great mass of bent and twisted tendrils wound this way and that, forming half a circle that stood approximately twenty feet in the air.

The tree seemed to simply lay down, lifting its roots out of the ground as if they were nothing more than a flat base on which it had stood. Some of the suspended roots ventured close to the path, and one even reached about waist-high out into the area that Marissa passed by.

The massive tree had obviously been down for some time, as the wood was rotted through in some spots. Marissa was dismayed that nothing else had grown in the area vacated, but she felt sure that new life was not permitted in this cold, evil forest.

Thinking of the cold, she shivered involuntarily. The temperature seemed to be dropping steadily the farther she went. The crystal grew warmer, radiating a heat as if already doing battle with the evil. She marched on, eager to commence, so it could be over and she could return to the warmth and love of her kingdom.

Up ahead a short distance, she could see the trail curve to the left. Since it had not had any such curves previously, she prepared herself, feeling she might finally be getting close. She reached the curve and slowed her pace. Not knowing what awaited her, she would not rush in blindly.

Moving cautiously along the path, she noticed the crystal had begun to pulse. It too seemed ready. She rounded the curve, and came to the beach of a vast body of water. The path ended at a line of sand that ran straight into the lake. The sand was brown and green, as if moldy and rotten, and spanned about fifty feet. She knew this would be her arena.

Marissa stepped onto the sand, and felt it give softly beneath her slippered feet. It shifted slightly, just as she expected sand to do. She looked out over the water. It was flat and still, with no breeze to cause a ripple. The water itself was menacing. It was black and evil-looking, not at all like what one would expect lake water to be. She proceeded to the edge, and looked into the depths.

Where sand met water, there was no movement. The water lay still, as if in a bowl, and even at the very edge, she could not see into it. Murky fell extremely short in describing this stagnant pool. It was completely black, totally opaque, and had a rancid smell emanating from it, making Marissa think of a pool of rancid petroleum. She could feel evil and death reaching out with icy fingers, as if to pull her down into the deepest pits of a black and oily grave.

She stood holding the staff before her, its tip resting on the sand. The crystal pulsed atop it, just an inch or so above her head. Taking a deep breath, Marissa called out over the water. "I am here!"

As her last word was spoken, a sudden wind whipped at her, startling her and making her gasp. The crystal flared in response, no longer pulsing, but maintaining a glow that reflected off the dead water. In her mind, she heard the reply of, "As am I."

She could see nothing that might be her enemy, but instead felt its presence. No shadows danced on the water, no beast rose from its depths, simply a presence, intangible yet filled with hate. She considered asking it to show itself, but then realized it had no earthly body to show.

She drew herself up to her full, if not quite impressive, height of five feet five inches, and stated, "I command you to cease your aggressions. You will leave the people of Avonridge in peace."

She could feel laughter, a low, vibrating chuckle.

"You overestimate your power," it seemed to say. But she was certain of the path she must take to battle this entity.

"I am Queen of this land," she continued. "Ruler of all within it." As she spoke, the crystal once again began to pulse, as if to bolster Marissa's words.

The wind blew once again, slashing around her, catching her hair and skirt, twisting and pulling, as if to rip them both from her body. "You do not rule me," the force said in her mind. "I am not a lowly subject."

But she could feel misgivings within those words, and so she plodded onward. "All who dwell within Avonridge are my subjects. It is the law of the land. You choose to dwell within Avonridge; therefore, you are my subject."

The wind whipped harder, angry fingers tearing at her clothes. "I am subjugated by no one," it roared through her mind.

But still, she continued. "As long as you dwell within the land of Avonridge, you are subjugated by me!" Her skirt flapped violently in the wind, her hair felt like hands were entangled in it, trying to rip it from her head. The gold tiara she wore seemed in danger of being flung out over the putrid lake, but somehow remained where it belonged.

Suddenly the wind stopped. For a moment Marissa thought it was over, but then she felt the voice speak very calmly to her.

"Under normal circumstances your words would be correct. However, I am different. You cannot control me. There is none other like me, and therefore I follow no one's laws."

She felt uncertain. Could that be true?

"If you want something from me," the force continued, "you must bargain for it."

"Bargain for it?" she said. "What kind of bargain?" She could feel great satisfaction from her enemy, and she grew more uncertain.

"You want me to leave Avonridge," it stated. "I require recompense for giving up my home."

"What kind of recompense?" she asked uneasily.

"Something very small," it said, and Marissa could feel it jeering. "Something very insignificant." It paused. She grew more uneasy as each moment passed. The being let it drag on long enough to cause her heart to grow cold.

"I require the lives of five hundred of your children. Each queen before you has paid the recompense, and I have kept the bargain. You will pay as well, and I will leave."

Could this be true, she wondered desperately. Could that be the way the previous queens had rid Avonridge of this evil? Was that why Camille wouldn't tell her any details, and simply stated she would know what to do? Fear gripped her heart at the idea of offering children to this evil.

"Five hundred babes must be sacrificed to me," it continued, "and I will leave for five hundred years."

She heard the words in her mind. The logic was there, five hundred lives for five hundred years. But suddenly the staff grew hot, searingly hot, burning her hand. Anger flooded Marissa's heart, dispelling her uncertainty. The crystal flared with luminescent power, and she knew the words for what they were; lies to try to defeat her.

The entity seemed to read her thoughts, and the wind returned with a vengeance. The first gust caught her squarely in the face, and she staggered back a step with the force. Knowing she was right gave her renewed courage and determination. She could feel the spirits of the dead queens surround her. The crystal continued to glow, but was changing as well. The heat it was giving off took on a solid quality. It expanded and grew, enveloping Marissa in a bubble of warmth. It completely engulfed her, and as soon as it did, the wind stopped.

She stood on the sand, looking out over the water. She was indeed within a bubble, a translucent bubble of protection. She glanced around and realized the wind had not actually stopped. She could see it buffeting the trees at the edge of the sand, and she could hear it whining past her, but she was completely untouched by it.

The crystal began to pulse again. It maintained the bubble, and Marissa was able to once again fight this enemy. "There will be no recompense," she declared. Then, in a quiet, commanding voice, "You will leave Avonridge because I command it. You dwell within my kingdom, and therefore you will obey my decrees simply because I command it."

The sound of the wind increased. Sand blew up in dirt devils to dance, and then be thrown across the beach. Dead sticks and twigs sailed through the air, one even bounced off the bubble as it flew straight at Marissa's head.

"I will not be subjugated," the entity roared. "I follow no one's laws."

"You will follow mine," Marissa said quietly, absolutely.

The entity seemed to be in a full-scale temper tantrum. The water that had remained calm and flat until now, suddenly surged. It raced up onto the shore at Marissa's feet, but when it encountered the bubble, it was turned aside to rush harmlessly past her as if she were an island in a stream. Sticks continued to fly, and even a rock began to roll along the sand.

"I will not, I will not!" she could feel it scream, and she was indeed put in mind of an unruly child. "I follow no one's laws!"

"You follow the laws of the sovereign of this land," she calmly replied. "And I command you to leave Avonridge."

Just as quickly as it had started, the tumult ceased. Flying sticks dropped from the air, the trees were no longer bending and moving, and the water once again lay flat and still.

The crystal abruptly went dark and the protective bubble disappeared. Marissa was once again standing on the calm shore, and it seemed as if nothing had occurred. She looked at the crystal, and it appeared almost dull in comparison to moments before. In fact, it no longer glowed at all, and became simply a pretty gem atop a staff. The staff was changing also, reverting to the brown color it had been before the crystal had been placed upon it.

Relief flooded Marissa so completely that her knees buckled. Grasping the staff to remain upright, she took a deep, trembling breath. She didn't need to ask if it was over, she could feel that it was.

THIRTY-SEVEN

As Marissa walked back through the Vulture's Claw, the air did not feel any less forbidding, any less hostile. But the source of the hostility was obviously absent, and so she walked with a lighter step. She was anxious to be free of the angry forest, and hurried along the path. She could see the edge of the trees, and was just about to step through when a feeling stabbed her stomach.

She stopped just inside of the forest, steps away from the field beyond, and waited, trying to read the feeling. It was a warning, silently cautioning her to continue to be vigilant. She couldn't imagine why her senses would so violently caution her about stepping free of the Vulture's Claw; wasn't the evil supposed to be contained within? Well, whatever the reason, her senses were screaming to her, and she would listen.

Slowly she walked to the end of the path at the edge of the trees. With each step she could feel her stomach knot up tighter. She continued carefully, looking all around, keeping her ears alert for any sound. She paused at the very edge of the forest. Off to the side, just inside the line of trees, snagged on a branch, was a piece of cloth. Marissa stepped up to it and studied it. It was green velvet, and she felt she knew it from somewhere, but couldn't put her finger on it. Thinking this anomaly was what must have given her the uneasy feeling, she prepared to exit the forest,

However, as she lifted her foot to step out, she heard it. She didn't recognize it at first. But then it sounded again. A growl, low and menacing. She listened intently. It sounded just like the growl of a dog, a very large dog. Then she heard another, similar to the first, but from a distinctly different dog. After that, a third. It now appeared there was a pack of dogs, or could it be wolves, just outside the Vulture's Claw.

Marissa had heard nothing about a pack of dogs wandering through Avonridge. Where would they have come from, and why would they be waiting outside the Vulture's Claw? She had always liked dogs, even though she was really a cat person, but she of course had great respect for the destructive capabilities of a pack.

She edged closer, painfully aware that her only weapon was a simple staff with what was now a purely ornamental crystal tip. She looked this way and that, finally spying the growling creatures. They were indeed wolves, a pack of wolves, waiting just outside the Vulture's Claw, waiting for her. To the left she could see three, all circling close to the trees, but obviously keeping a safe distance. To the right were two more. Five. Five wolves waiting for her.

A vision of evil flashed through her mind, not the all-consuming evil of the Vulture's Claw, but an evil tinged with hatred and anger. The stooped, thin image of Marcellus formed in her mind, and she believed she knew how the wolves came to be there. A white-hot rage burst through her body as she realized the wolves had been purposely set upon her for the sole intent of stealing her kingdom. She gripped the staff, and forced the rage back down, knowing it would be a hindrance to clear thinking.

The wolves stopped circling and all five turned to face her. Their eyes were bright and, their huge jaws hung open, tongues lolling like relaxed panting dogs. For a split second they looked like large, friendly family pets. Then, in unison, their jaws closed and all five growled. The family pets turned instantly into a savage hunting party.

Marissa waited, and so did the wolves. They did not advance, and she realized they would not step into the Vulture's Claw. The irony was not lost as she understood that the evil she had fought to rid her kingdom of unnatural death was what was keeping these beasts from attacking and ripping her apart. Her mind flew over strategies of escape, but kept returning to the unmistakable fact that she faced five killing machines with only a stick as a weapon.

They stared for a moment more, then resumed circling, keeping their eyes on her as they paced. It was clear that as long as she stayed within the Vulture's Claw, they would not attack. Unfortunately, she couldn't stay here forever. She lifted the staff, feeling a mixture of emotions running through her, not the least of which was fear. The fear angered her. Since she had become Queen of Avonridge, she had been able to banish her fears. But now, without the Dragon Slayer, she felt vulnerable, a feeling very unfamiliar lately, and one she abhorred and resented.

Anger churned and bubbled within her stomach as she watched the wolves watch her. She kept the rage at bay, and forced the fear to the back of her mind, refusing it a toehold. With the staff in the air before her, she took a deep breath, and attacked.

She leapt from the Vulture's Claw, the skirt of her white gown billowing, and swung the staff like a club. She connected with the massive head of the one she believed was the lead wolf, and, to her surprise, he was knocked about ten feet to the side with a pained yelp. She immediately brought the staff back around and struck another as he came at her. This one was also knocked down, and the remaining three hesitated.

She used this to her advantage and advanced, staff raised menacingly. The three backed away, however the first wolf had risen and was slinking closer, anger evident in his yellow canine eyes. His head was slung low between his hunched shoulders, and the gray and brown of his coat seemed to magnify his anger.

The second wolf was on his feet again also, and so the unscathed three regained their courage. The five formed a half-circle around Marissa, and she was reminded of her vulnerability. She held tight to the raised staff, ready to swing at the first attacker to make a move. The wolves held their positions, but continued to pace a step this way and a step that way, like a continually moving wave before her.

Deciding to be on the offense instead of the defense, she raised the staff and prepared to attack again. Just as she moved to step into her assault, a screech of rage pierced her ears, and a little black and white ball of hissing, slashing fury sailed right onto the huge head of the lead wolf. Claws and teeth ripped and tore at the muzzle of the wolf, who immediately yelped and howled in surprise and pain. The wolf ran backward, shaking his head in a desperate attempt to rid himself of the enraged Merlin, but the hissing cat held tight, slashing and biting.

The other four wolves jumped back, obviously unsure what to do. Marissa again swung the staff at the wolf she had previously felled, and he was knocked to the ground again. He quickly rose and backed farther away as Merlin continued his furious assault. The screeching of a battling cat and the yowling of a wolf in pain reverberated through the air in a deafening crescendo. The three wolves that had managed to keep out of the melee finally decided enough was enough, and turned and dashed across the field in frenzied retreat.

Marissa stood facing the second wolf, who was obviously thinking better of this planned attack on the queen. He glanced at the disappearing hind ends of his comrades, looked past Marissa at the lead wolf being mercilessly mauled by a hissing, spitting ball of claws and teeth, and allowed discretion to be the better part of valor. He spun to his right and raced after his fleeing buddies.

Marissa watched him for a few seconds to be sure he was truly going, then turned to the lead wolf and Merlin. She stared in amazement at the little spitfire as he raked and bit the beast's head. The wolf was spinning and tossing his head, trying to dislodge his assailant, but Merlin held firm, determined to teach this brute a lesson. The wolf howled, throwing his head to the ground and raking his front paw across his brow, but still the cat held on. Blood was splattered everywhere; on the wolf's gray coat, on Merlin's black and white fur, and dribbled over the ground.

Finally, when there was no question as to the victor, Marissa called softly, "Merlin, let him go." The cat leapt into the air and landed a few feet from his queen. Finding himself finally free of his attacker, the wolf spun and raced away.

* * *

Marissa stalked into the palace with Merlin at her heels. She was greeted by Sir Erick, Ruth and Roland, all gushing thankful greetings at her for returning alive. But when they saw the anger and determination upon her face, and the blood on Merlin's coat, they became silent.

"Majesty," Ruth said, rushing up to her. "Are you hurt?"

"No. Bring Marcellus to me," she demanded without further explanation, and walked straight to her chambers. All three stared after her, then Ruth hurried to follow the queen, and the Black Knight and the Captain of the Guard went in search of the old magician.

Marissa entered her chambers, walked to the far corner and stood the staff against the wall. Merlin found a patch of floor away from any furniture, and began to clean his coat.

Ruth stood in the doorway, unsure exactly what to do. "Majesty," she called softly. "Did it not go well?"

Marissa was angrily pacing, paying little attention to the concerned woman. "It went as expected," she replied shortly, but offered nothing more.

Ruth went to the sideboard and filled a basin with water from the white pitcher beside it. She took a cloth from a drawer, dabbed it in the water, and knelt beside Merlin, wiping some of the blood off his coat. He seemed thankful for the help, and allowed her to clean him, purring as she rubbed and blotted.

"Majesty," she said again. "Is Merlin hurt? There is blood but I do not see a wound."

Finally, Marissa stopped and looked at her. "No," she said. "Neither of us is hurt."

"But the blood?" Ruth said.

She shook her head distractedly. "It belonged to the wolf."

"Wolf? What wolf?"

Marissa looked at the distressed young woman, took a breath, and softened, realizing Ruth's concern. Not wanting to worry her more, she downplayed it. "It is over now. Nothing to bother yourself about."

Ruth looked at her, obviously wanting to pursue it. But said no more. She continued to clean Merlin, fussing over the cat. Suddenly Sir Erick and Roland appeared in the doorway.

"Majesty," the Captain of the Guard called. "We have brought you Marcellus."

Marissa faced the two men that stood side by side in the door, and spied the hood of the old man's cloak behind them. They stepped aside to reveal the scowling countenance of the one she was sure had tried again to disrupt her kingdom, possibly even to kill her.

"So, Marcellus," she called. "It seems we have a matter to discuss."

His expression didn't change as he glared at her. "I am sure I do not know to what you are referring," he sneered.

"Please, come in," she said, her voice an arctic command.

Slowly, but without the slightest reservation, he hobbled into the Queen's chambers. Ruth stood up and, after glancing at Marissa, hurried out the door. Merlin issued a low growl from deep within his throat, but was ignored. Sir Erick and Roland stood in the doorway, obviously not wanting to leave their Queen alone with this threatening figure, but she dismissed them.

"Please leave us, gentlemen," she said. "Marcellus and I have business to discuss." They looked at each other uneasily, then obeyed.

Marcellus stood in the middle of the room, leaning heavily on his gnarled staff. He stared defiantly at the queen, but said nothing. She met his gaze, eyes blazing with anger.

"So, first you try to ignite a war between the High King and Avonridge," she began, "and now you call forth a pack of wolves to do your bidding. Are you really that cowardly as to hide in the shadows while you try to get others to do your dirty work?"

At the word 'cowardly', his countenance changed. He raised his eyebrows slightly in anger, but then appeared amused. He chuckled softly, sending a chill up Marissa's spine. "My dear lady," he said, with only a hint of sarcasm, "I am at a complete loss in regard to your accusations. I admit, it would have been most enjoyable to be afforded the opportunity to witness the usurper's defeat of an upstart such as yourself. However, I have absolutely no knowledge of a pack of wolves."

Marissa was caught off guard. Was he actually admitting to his attempt at manipulating Marissa and King Arthur into battle? It certainly seemed so. And so why would he then deny being involved with the pack of wolves? Perhaps to throw her off? It was possible.

"So, you admit trying to start that battle?" she asked angrily.

"If an admission would please you," he said haughtily, "then you have one."

"And the wolves, you set a pack of wolves on me," she accused.

He seemed truly unaware of this new situation. "I assure you, I have no such wolf pack. I do not even know of one."

She stared at him, trying to read his thoughts. Frustrated, she began pacing, as Merlin sat like a statue several feet away, growling menacingly. Marcellus waited patiently, obviously not intimidated in the least. She stopped and faced him again, still uncertain.

"Why should I believe you?" she asked softly. "You want my throne. Why should I trust anything you say?"

He pondered her question for a few moments, then said, "Your concern is certainly valid. It would be imprudent of you to simply accept my denial at face value. After all, I could say anything and you have no way of telling what may be real or what may be fabricated."

This further stumped and annoyed Marissa. She had been thinking that exact thing, but to hear him offer it made no sense. Why would he encourage her to doubt him? Or, at the very least, question his words? As much as she wanted to blame him for the wolves, she was beginning to believe he had nothing to do with them.

"May I offer this, dear lady?" Marcellus said. "I have said before, you trust the wrong people." She glared at him, but he put up a hand to quiet her. "Do not misunderstand, I have no delusions that you would trust me. Honestly, I do not care one way or the other if you do. However, you would do well to question those you consider loyal, perhaps even friends. Not everyone is as happy with you on the throne as you might believe."

Anger still churned within her breast, but she nodded at the old man. "I am quite aware that some still consider me an outsider here. But, if you are referring to those within my tight-knit circle, you are again just trying to cause dissention. I have complete faith in Ruth, Sir Erick and Roland."

"I am sure you do," he answered lightly. "Now, if we are finished here, may I go?"

His arrogance infuriated her, but she waved a hand at him dismissively. As he hobbled out the door and disappeared, she turned to Merlin. "How could I possibly doubt Ruth and Sir Erick?" she asked the little cat as she began to pace. "They're the ones who brought me here. They would have had any number of opportunities to prevent me from assuming the throne. And Roland, he is a soldier, a loyal soldier. I even get the impression he does not particularly care who sits upon the throne, his loyalties are with the crown no matter who wears it. I cannot even consider doubting any one of them, can I?"

"No." Marissa stopped pacing and looked at the little black and white creature who sat watching her. Had she heard correctly? Had he spoken, or did it just sound like it? A meow could sound like a word, but she'd suspected him of speaking before.

"Merlin," she said. "You are a complete enigma. I know you have intelligence beyond a normal animal, but can you speak as well? I would like to know exactly what is going on with you. Others seem to know, but no one wants to explain you to me. Would you please do so?" she asked, exasperated.

The little cat simply looked at her for a moment, then resumed his grooming.

"Oh!" she said, stamping a foot.

THIRTY-EIGHT

Ruth once again tried to begin preparations for yet another celebration, but Marissa immediately put a stop to them. She wanted no celebration, and in fact, forbade anyone from speaking of the Vulture's Claw again. She never wanted to hear it mentioned, either by name or simple indirect reference. It was time, once again, to put it away.

Life at Avonridge again became peaceful and mundane. Little Kevin was a happy, outgoing child, and Marissa fully enjoyed having him around. She spent as much time with him as possible, since the kingdom was running smoothly, and she was needed only occasionally for queenly duties. Approaching three years old, he still had not spoken a word, and Marissa was finally becoming concerned. The trauma of losing his family so violently could very well affect him for the rest of his life, but she was no child psychologist. She had no idea what to do, so she surrounded him with love, and hoped for the best. If she couldn't bring him out of his silence, she could at least make sure he had every other advantage possible.

The first snow of the winter arrived a few weeks before Christmas. Marissa loved the winter and bundled Kevin and herself up in warm clothes, and went outside to enjoy nature's playground. Just on the outskirts of Avonridge proper was a small hill, just high and steep enough for a toddler to enjoy. The village craftsmen were well prepared, and a small toboggan-like sled was brought for them.

Rather than make Kevin walk through the foot deep snow, she had him sit on the sled as she pulled it along. Immediately he started to laugh with delight, and as they began to climb the hill, he started to screech in anticipation. At the top of the hill, Marissa positioned the sled, climbed behind Kevin, grasped him to her chest, and kicked off. The sled moved smoothly, descending the hill, gaining speed as it went. Snow kicked up, spraying them both, and they laughed as they whooshed along.

At the bottom, they skimmed a small mound, and the sled tipped over, dumping the duo face first into the snow. They rolled over, laughing, and Kevin screeched, "More!"

Marissa stopped laughing and looked at the little cherub covered in white. He was still screeching, kicking his feet and grasping handfuls of snow. He looked at her and repeated, "More!"

She started laughing again, lifted him up and tossed him into the air. As she caught him, she agreed, "More!" and back up the hill they went.

Little Kevin had found his tongue, and now talked non-stop. Starting with the halting words here and there of beginning speech, he quickly advanced to full sentences. It seemed as if he kept up a steady barrage of questions about everything, from why is the snow so cold to why is Marissa's hair brown. She delighted in his inquisitive nature, and answered every question as honestly as possible. He learned quickly and truly loved acquiring knowledge of every kind.

The incident outside the Vulture's Claw with the wolves was all but forgotten as Marissa had heard nothing about the pack since. It almost seemed as if they had been conjured just for that particular purpose, and then sent away when it was done. She again remembered how lucky she had been that Merlin showed up when he did, and she didn't miss an opportunity to thank him with pets and tidbits of tasty meats. She truly owed a great deal to those that stood beside her in this wonderful, yet wild land.

Since Kevin was speaking now, and had the courage of innocence, she decided it was never too early to learn the things that would help him through his life. He would run through the palace, exploring everything he could, laughing an angel's laugh when he discovered something new, and so she gave the ladies-in-waiting each a different teaching task. Since she had made it clear that Kevin would be staying, and that she would tolerate absolutely no disrespect toward him, they had eased up somewhat on their dislike of the little ball of energy. In fact, the youngest, Louisa, even seemed to enjoy his company, that is when the others weren't around to tsk-tsk at her.

Just as Marissa had suspected, Kevin was a quick learner. He could count to ten and say the alphabet in no time. She smiled at him when he would recite in a sing-song voice the a-b-c song. She could see an intelligence in him that made her heart swell.

She was not sure where he'd picked it up, but one day, out of the blue, he called her Mama. She looked into his huge brown eyes and what she saw made tears well up in her own. He was looking at her the way she had seen children look at their own mothers; with complete trust and love. She hugged him so tight he grunted at her and tried to push away. Laughing, she set him down, and he again said, "Mama." From that moment on, she no longer felt he was her ward, she had become his mother, and he was her son.

Kevin's learning sessions continued until Christmas day. The kingdom was filled with the joy of the season, and Marissa was eager to have him see the Christmas tree and all the presents. The tree had been put up in the banquet hall and decorated on Christmas Eve after Kevin had gone to bed, and so was a total surprise.

She was waiting for him when Louisa brought him into the hall, and when he spied the tree with lit candles on the branches, and stacks of presents all around it, he screeched with excitement. Pushing out of Louisa's arms, he raced to the tree, but Marissa caught him and scooped him up in a bear-hug, whispering, "Merry Christmas, little one." She kissed his cheek, then released him so he could get to the bounty that awaited him.

Having no wrapping paper, Kevin didn't have to bother with ripping it off the presents, he simply started grabbing the wooden horses and cows and sheep, making them clomp on the hard stone floor as he pretended they were galloping around. Quickly tiring of the smaller playthings, he spied a rocking horse off to the side, and clambered onto its back. He grasped the handles attached to either side of its head and furiously rocked back and forth so hard Marissa was afraid he'd send it over backward.

The rapt joy on the little boy's face was the only Christmas present Marissa needed. She spent hours watching Kevin giggle and laugh as he experienced all the wonderful gifts the woodworkers had created. Besides the rocking horse, there was a shoo-fly, a wooden seat inside a frame that swung forward and backward without the possibility of flipping over like the rocking horse could. Obviously, this had been made by a woodworker who had children.

A small cart stood next to the Christmas tree. This was for a pony to pull, and was sized for Kevin to last at least a few years. Beside the cart was a small wooden boat, but Marissa could not imagine letting him captain it alone for quite some time. Even so, he climbed into it, instinctively standing in the bow with one hand raised as if leading a naval fleet into action.

Late in the afternoon, after Kevin finally exhausted himself and had begrudgingly taken his nap, Marissa donned her gleaming white gown, placed her purple crown atop her head and threw her purple cape over her shoulders. The Christmas mass was about to begin, and she would not be late. Grasping Kevin's tiny hand, she escorted the little tot out the massive doors of the palace and down the steps. He looked adorable in his white ruffled shirt and scarlet waistcoat and tights, as he marched beside her, as solemn as a mortician.

Ruth and Sir Erick followed close behind, with the gaggle of ladies-in-waiting bringing up the rear. Even on this most blessed day they continued their sniping at one another, as this one hated that one's gown, and that one thought this one's cloak too gaudy. Marissa refused to let them dampen her spirits, and so tried her best to ignore them.

Sitting in the front pew of the church, Marissa marveled at how well Kevin behaved. He seemed to understand the importance of decorum, and was silent throughout the entire service. It felt to Marissa that it took all afternoon. The incense tickled her nose and caught in her throat, and she struggled to keep from sneezing and coughing. Finally, the bishop made the sign of the cross over the congregation, and Marissa stood up, feeling guilty that she was so anxious to be going.

She did, however, have a reason to be anxious. She would be greeting many of her subjects outside the church and handing out Christmas gifts to the boys and girls. She would also be giving the parents a Christmas goose, a little too late to be prepared today, but just as good for tomorrow.

Months ago, she had set the village craftsmen to making hundreds of dolls and toy soldiers for the children of Avonridge. They had risen to the task marvelously, and she was looking forward to handing them out. Knowing that simple existence was often a struggle, many of her subjects just couldn't afford to give their children non-necessities like toys. She was happy to be able to do that for all the wonderful people of her kingdom.

As she handed dolls and toy soldiers to the children filing past, Ruth and Sir Erick took care of the distribution of the Christmas geese. The winter may just be starting, but Marissa was determined not to let even one of her subjects go hungry.

As the day waned, Marissa grew as sleepy as Kevin. She looked forward to climbing into her bed and allowing all the happy images of her first Christmas in Avonridge to replay all through the night. When she finally pulled the covers up to her chin, she smiled, feeling more satisfied than she could ever remember.

THIRTY-NINE

The day after Christmas put things back to normal. Kevin's lessons resumed, and Marissa began seeing to palace activity. She once again found the room with the harpsichord and called on Wanda, the lady-in-waiting who played, to teach her. For the first lesson, they spent an hour, and Marissa was happy to find a common bond with the woman. She still did not feel completely comfortable with the ladies, but at least now they were serving some purpose.

When her music lesson was over, she went in search of Kevin to bring him to dinner. As she walked into the room where Bethany was giving him his decorum lesson, however, she was struck with fear by what she saw.

Bethany was sitting in a corner of the room in a large chair, chin on chest, snoring softly, and Kevin sat on the floor in the middle of the room with Marcellus. As her eyes fell upon the old man and she could see his close proximity to her darling boy, her stomach gave a lurch and she thought she might be ill. They did not seem to know she was there, and for several moments she could not find her voice to stop whatever they were doing.

Unable to do much of anything else, she watched the two for a while. Marcellus held a small stone in the palm of his left hand, waved his right over it, and Marissa watched as the stone levitated several inches into the air. As the stone dropped, she was just about to force a word or two but stopped when Kevin brought his little hand up to mimic the old man.

He held his hand over the stone as Marcellus had done, and waved it in the same way. As Marissa watched, the stone responded exactly as it had for the old magician; it levitated. Kevin laughed excitedly, and Marissa gasped. At the sound, both the boy and the old man turned to face her. Kevin was still laughing, and Marcellus' eyes were crinkled in a smile.

"Ah, Majesty," he said easily. "I see you have witnessed the boy's talent."

She dragged her eyes from the happy face of the student to the veiled expression of the teacher, for she realized this is exactly what was transpiring. Marcellus was teaching Kevin the ways of the magician, and it seemed apparent that Kevin indeed had a proficiency in the area. She shook her head, not wanting to believe what she had just witnessed.

Marcellus kept his thoughts closed off, not allowing Marissa to know exactly what his intentions were, and all manner of horrid motives flashed through her mind. Was he trying to corrupt the boy? Would he try to influence him enough to steal him away from her? She shuddered with each new thought.

Finally, she found her tongue. "Explain this!" she demanded.

Marcellus chuckled and replied, "Of course, Majesty." He rose from where he had been sitting beside Kevin, and stood up stiffly. He lazily stretched the aches out of his muscles, and Marissa grew more impatient with each passing second that he did not answer. She could see that he was enjoying her frustration, so she waited for him to continue.

"It seems," he finally said, "that little Kevin here, has a very real talent. As you just witnessed, the magic is strong within him. He will make an excellent apprentice."

"Apprentice," she echoed. "He will be no apprentice of yours," she stated firmly. "You come in here uninvited, put my lady-in-waiting under some sort of spell and take advantage of the situation. I would not trust Kevin in your company, he is too young to defend himself."

"There is no need for defense against me, I simply wish to help the boy achieve his potential. And as for your lady-in-waiting, she is not under a spell, she simply accepted the suggestion that she…take a nap." He smiled slyly.

"Well, Kevin will not be your apprentice," Marissa repeated.

"Ah, Majesty, do not decide so hastily. You do not fully understand the magnitude of what is transpiring here."

"I understand that you will not be allowed to corrupt this innocent little boy."

"Corrupt?" he said, genuine surprise on his wrinkled old face. "The mastering of the talent is by no means corruption. He will grow and flourish with its use, and the power it brings him."

"The way you've grown and flourished?" she shot back. "The way you use your talent to deceive and trick people? Is this what I should allow for Kevin, to become a traitor to the rightful sovereign?"

Marcellus was silent for a few moments. "Yes, I understand you feel I am a traitor to the rightful sovereign. I had indeed intended to acquire the crown for myself."

"Had intended?" she said. "And you would have me believe you no longer intend?"

"Circumstances have changed," he replied. "I was sure you would fail in at least one of your challenges and in the ensuing pandemonium, I could easily step into the position. However, you disappointed me at every turn. You were victorious each time you were called upon."

"And that frustrates and angers you," she challenged.

"Yes, it does," he stated, eyeing her with cold arrogance. "I was fully expecting an easy mark. I suppose I should have trusted Gertrude's instincts more, but she could be an old fool at times. I expected her wisdom to falter, but it did not."

"I am sorry to be such a disappointment to you," she said icily.

"Oh, I know you indeed are not sorry, but that is neither here nor there," he easily replied. "You have accepted your position with the ability needed to rule Avonridge. I have no more designs on the throne at the moment. I will be quite happy with my new apprentice. But rest assured, should the time come when you do fail in your challenges, I still fully intend to step in and assume the throne. In the meantime, however, little Kevin will keep me quite busy."

"No, he will not," Marissa stated. "As I said, he will not be an apprentice to your dark magic."

Marcellus nodded slowly, absorbing what the queen had just said. "Dark magic," he repeated thoughtfully. "Yes, I suppose my magic could be considered dark. But it is only in the manner in which I yield it that it would be seen that way. The magic is neither good nor dark, it simply takes on the characteristics of its master."

"An excellent argument against you teaching him," Marissa said. "Your ways are dark, your ideas and desires are dark, and I will not have you influencing this child with your dark designs."

"I have no desire to concentrate on the dark aspects of life," he stated. "Some time ago I had chosen anger as my way of walking this world. The child may choose his own way. In fact, his very presence eases the anger in my soul. With an innocent life to guide and teach, my life would be greatly improved. So, his apprenticeship would be beneficial to us both."

She was silent, so he continued. "The talent he has been given is immense. It is a gift to be honed and used. Why do you suppose he was spared from the fire that claimed his whole family?"

Again, she shook her head, wondering just how true any of what he'd said could be. She was the one person in the kingdom who was not easily fooled by him. Others could be influenced by his power, as Bethany obviously was, but she could not, and now she was finding herself believing this man.

Had he indeed abandoned his plans to openly try to take the kingdom from her? She was beginning to believe exactly that. But would his attention now be diverted to Kevin? She had to admit it was a miracle that he'd survived the fire. Could it be that he was meant for bigger and better things? She considered that, but still did not trust Marcellus.

"So now you are no longer trying to take the kingdom from me, you will settle for taking my ward," she challenged.

Again, he chuckled. "There is no taking to be done," he replied. "I can simply tutor him as the ladies-in-waiting are doing. Of course, my lessons are infinitely more important than whatever those," he paused, "women may teach him," and Marissa heard unmistakable contempt in his voice.

"You feel the ladies-in-waiting are not up to the task of training Kevin in the lessons necessary for life in the kingdom?"

"I feel the ladies-in-waiting are more suited for training barnyard hens, as that is exactly what they are."

Marissa almost chuckled herself. It was extremely strange to so completely agree with this man on anything, but she did in fact agree.

Bethany, who had been sleeping quietly in the corner, snorted and awoke. She glanced at Marissa, then Marcellus and finally down at Kevin. She immediately realized that she had fallen asleep and left her charge unattended, and the panic she felt was obvious.

"Majesty!" she screeched breathlessly as she stood up quickly. "I was only, I mean there was no one, I only just…"

Marissa put up a hand to silence her. "Enough," she said, and Bethany was silent. Marissa wouldn't be too hard on the young woman, since Marcellus was mainly to blame for her dozing off. But she also couldn't just let it go, after all, Bethany was supposed to be caring for Kevin. "I will deal with you later," she said darkly. "Now go." And Bethany ran from the room, a hand over her mouth to muffle her sobs of anguish. Now Marissa did laugh softly.

Kevin, who had been sitting on the floor playing with the stone, turned to Marissa and, smiling, said, "Mama, see what I do."

She walked to him as he levitated the stone again. "Yes, I see what you can do. It's wonderful." She crouched down and tussled his hair affectionately.

"Majesty," Marcellus said. "It would be to his advantage to utilize the talent he has been given. To waste it would be a shame."

Without turning to look at him, she said, "I still do not trust you. I have no way of being sure you aren't planning some heinous trick that would devastate me or my kingdom."

"Very rarely does a talent such as this manifest. The fact that it has in a circumstance where I may be useful is a great honor to me. Teaching this child to use his talent would humble me. I have no ulterior motive."

She was silent for several moments, still refusing to look at him. "This is something I must consider. I will let you know what my decision is. Until then, I expect you to stay away from my ward. He has become a son to me, and if you use trickery to approach him again, I will consider it an act of aggression, and take the necessary action."

"I understand," he said, only a hint of anger in his voice, and he turned and hobbled out the door.

The days following Kevin's "lesson" in magic were a struggle for Marissa. The stone he levitated soon became inadequate, and he began levitating spoons, bowls, goblets, and Marissa's struggle was to keep him from performing for an audience. She wanted no witnesses to his talent. Should anyone learn of his ability, she could only imagine the problems it would cause. The ladies-in-waiting were only just getting over their disdain of him, and she didn't need anyone deciding he was possessed.

Knowing he would use his gift as often as he could, she had no choice but to keep him with her at all times. She was constantly taking things away from him just as it was about to be raised off his palm.

Finally, being able to neglect her duties no longer, she decided to let Ruth in on the secret. Ruth would be able to keep Kevin in Marissa's chambers and let him practice to his heart's content. Unfortunately, Ruth was not at all happy with her new assignment. She was obviously frightened by Kevin's ability, but tried to hide her fear for her queen. Every few hours, Marissa would go to her chambers to check on them, and Ruth would be in a chair in a far corner, as far away from Kevin and his magic as possible. Marissa understood just how uncomfortable Ruth was, but could figure no other way to keep him out of the eyes of everyone else in the castle.

She was mulling over the decision she knew she had to make, but it became no easier with the passing of the days. She did not, would not, feel comfortable leaving Kevin in Marcellus' charge. Besides, she wasn't sure exactly how it was supposed to be done. Would Kevin go to stay with Marcellus? Would Marcellus take up residence in the palace? Just what was supposed to take place? Having had no previous experience in the matter, she simply didn't know how an apprenticeship was supposed to work.

Well, she was sure of one thing; Kevin would definitely NOT be going to live with the old man. Any tutoring would be done within the safety of the palace, where Marissa would at least be close by. She began to figure out something of a game plan. Marcellus could begin his lessons in the morning right after breakfast, when Kevin was rested, well-fed and alert. Then, after lunch, the ladies-in-waiting could resume their lessons. Hopefully, by then Kevin would have had enough practice to be satisfied for a while, and not try to levitate the ladies.

With the details fixed, Marissa realized she had in fact made her decision. And so, one morning, a week or so after she'd discovered Marcellus with Kevin, she sent for the old man. He hobbled up to her as she sat on her throne, and said, "Majesty, you sent for me?"

She kept her thoughts hidden from him, making him wait while she took her time in responding. Finally, she said, "I have been thinking long and hard about our conversation."

He remained quiet, waiting for her to continue.

"It seems some of your arguments are convincing. There must be some reason why Kevin was spared the same fate as his family, and it certainly seems the magic had some influence in that miracle. To ignore that influence would be to ignore a large part of what Kevin is. He obviously has some great talent, and it would be a shame to waste it."

Marcellus' expression didn't change as she spoke.

"And so, after much deliberation, I have decided that you may in fact tutor Kevin." Still his expression remained unchanged, and as he nodded in agreement, Marissa knew he was not in the least surprised by the decision.

"Very good, Majesty," he said. "When my I take him with me?"

"Oh, you will not be taking him anywhere," she said.

"Then exactly how am I to tutor him?" he asked, his voice hardening slightly in challenge.

"Exactly as you showed him his first little trick," she replied. "You will come to the palace each morning and give him his lesson. This way, he will still be in an environment where he is loved and will feel safe. His home and family need not be taken from him a second time."

Marcellus nodded again, but it was obvious he was not completely happy with the situation. However, he knew Marissa would not give in on this point, and it was better to have a few hours each day than nothing at all.

"Very well, Majesty," he said. "As you wish. I will begin his lessons first thing in the morning." And he turned and hobbled out of the room.

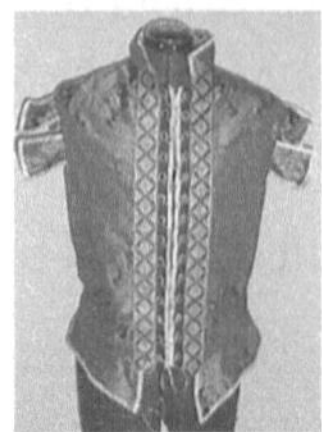

FORTY

With Christmas past and time marching ever forward, the kingdom of Avonridge once again returned to normal. Daily life resumed and everyone at Castle Avonridge went about their duties, ensuring the smooth running of the kingdom.

To Marissa's annoyance, Forsythe was coming round more often. His first holiday visit was just after the new year, just about the time Marissa gave Marcellus permission to tutor Kevin.

Forsythe caught Marissa alone in a corridor, and she wished she could have disappeared. Oh, he was his usual charming self, again presenting her with a rose. Unfortunately, he changed the topic of conversation from how beautiful the snow looked blanketing the kingdom, to how lonely she must be with no one close to her station with whom to converse.

"Be assured," she said, "I am not in the least lonely."

"I am sure you are kept busy," he quickly replied. "But how fulfilling can it be when, after carrying the burden of ruling the kingdom for the day, you have no one with whom to share that burden."

"That's actually somewhat funny," she said. "I have yet to think of my duties of ruling the kingdom as a burden. I am quite happy with my lot in life. In fact, just a few short months ago I was indeed lonely. My returning to Avonridge has been nothing short of enchanting."

"Yes, the kingdom is an enchanting place," he agreed, but Marissa could hear a cold edge to his voice. "However, you miss my point. I am simply acknowledging the fact that you have quite the large duty to perform all by yourself. It would be easier on you if you had someone by your side who was not of a lower station."

Marissa raised an eyebrow at him. "And who within the kingdom is not of a lower station than me?" she asked, forcing the note of challenge down as far as she could.

As Forsythe walked, his hands were clasped behind his back, and now he dipped his head and chuckled in a self-deprecating way. "I do not mean to appear boastful," he said, "but my lineage is of the highest standard. My forefathers have been dukes, earls, and even a prince or two."

"Most impressive," Marissa acknowledged sweetly. "But with the dukes, earls and a prince or two, has there been a king thrown in anywhere?" Her voice held not a bit of disdain, but the meaning was clear.

Forsythe straightened. "No, Majesty," he said, and the contempt was clear. "There have been no kings."

"Pity," she said softly. They walked on in silence for a few more steps, then he seemed to rein in his anger, and tried again.

"I am simply trying to be a good and dutiful subject," he said. "I see my queen obviously in need of a helpmate, and I am offering a more than acceptable option. It must be overwhelming to be expected to act completely on your own."

"That is a noble gesture," she said, and tried to give a little benefit of the doubt. He was arrogant and full of himself, but he wasn't exactly being disrespectful; chauvinistic yes, but not disrespectful. If, in his warped sense of reality, he honestly believed he was offering an acceptable option, well she wouldn't fault him completely.

"I appreciate the offer," she said. "However, I am not in the least overwhelmed. I am quite confident in my own abilities…"

"Please do not misunderstand," he cut in. "I am not implying that you are lacking. Only that you could benefit from some help."

"As I was about to say," she continued, "my own abilities are far from lacking, and I have the assistance of people I trust. I have no burden, and my duties give me pleasure and satisfaction. I have no need for anything more."

"I understand, Majesty," he said, and she again heard contempt as he spoke her title. "Well, I have enjoyed our little visit, but I have matters to attend to myself. I beg your leave," and he turned to go.

"Forsythe," she said, calling him back. He stopped and looked at her questioningly. "Have you forgotten something?" she asked.

"Forgotten something," he repeated. "Pray, what have I forgotten?"

"It is customary for royal subjects to show respect when encountering and leaving the monarch, is it not?"

His face turned a deep shade of purple, and Marissa recognized it not as embarrassment, but rather anger. He did acknowledge the statement, however, and said, "Of course, Majesty. Please forgive my foolishness, it was but a mere oversight." And he theatrically threw his arm out to the side, brought it across his waist and bowed deeply with a flourish. He then stood, turned on his heel and strode away.

She watched him go, fully understanding the insult, and decided she would no longer give him the benefit of the doubt. As his scarlet waistcoat disappeared around a corner, she had a nagging feeling that she couldn't place. Oh well, it was probably just from this encounter. She really couldn't understand why he would try so hard at something he must realize was a hopeless cause. But she guessed he probably didn't realize it was hopeless, or he would have given up already.

Little Kevin's lessons were continuing nicely. He was doing better with his letters and words, with his numbers, oh, and that other thing. He was able to levitate chairs and small tables now, and he obviously loved it. Marissa had been able to impress upon him that he was not to perform his levitation in the company of anyone but her and Marcellus. He seemed to understand, and so far, had complied. It also seemed to be a plus that he knew he would be given ample opportunity to do his little tricks during his lessons with the old man.

As Marissa had expected, the magic lessons soon progressed to other feats. Unfortunately, the other manifestations of his talent were less easily controlled. He quickly learned how to cloak himself so others could not see him, and was beginning to 'disappear' from his other lessons. Once he even 'suggested' Louisa take a nap, and while she dozed, he wandered out of the room and through the palace. Fortunately for Marissa, she was immune to Kevin's tricks just as she was to Marcellus'. Kevin had to learn the hard way, however, and spent fifteen minutes sitting in a corner for trying to get Marissa to think he had disappeared.

If she was to admit it, however, she was impressed with his ability. He obviously had a great propensity for what Marcellus was teaching him, and he learned quickly. For his part, Kevin really seemed to enjoy learning his magic, much more than his other lessons, and if she thought about it, she couldn't much blame him. She even found herself wondering what it must be like to be able to wield magic whenever you wished. She wistfully thought about the idea of being able to disappear when Forsythe came looking for her. Wouldn't it be lovely to see him approaching down the hall and simply will him not to see her?

But, alas, she possessed no such talent. She would have to stop worrying about being rude and deal with him appropriately. After all, he was rude and arrogant. Perhaps that was the only behavior he would understand.

It was a shame about him anyway. He could be so charming, but Marissa realized that was all an act. He was trying to achieve something, and he obviously didn't care how he went about accomplishing it. She remembered back to her arrival banquet. When Forsythe first approached her to dance, everyone seemed on edge; Ruth, Sir Erick, even Queen Gertrude. They all watched closely the entire time.

She wanted to get some more information on Forsythe, and so went in search of Ruth. She found her in the kitchen putting the last touches on the plans for the evening meal, and the two women walked down the corridor together.

"I wanted to ask you," Marissa began, "about Forsythe." She felt Ruth bristle at the name.

"What is it you wish to ask, Majesty?" she said, the dislike of the man evident in her voice.

"I first met him at my arrival banquet, and when I danced with him, it was clear that no one was happy about it. Everyone got very uptight about my being with him."

"Oh, really?" Ruth said with false curiosity. "What makes you say that?"

"Don't act like you don't know what I am talking about," Marissa said with only a touch of annoyance. "I saw the looks on everyone's faces when I danced with him. I could even see that Sir Erick was not pleased. Now, I would like to know why."

Ruth hesitated a moment, then said, "Very well, Majesty. Everyone was anxious because Forsythe is a manipulator. He has unreasonable aspirations, and he is not afraid to pursue them."

Marissa nodded. "Yes, I've noticed. But he could not have made his desire to marry me known before I arrived, could he? I mean, would he have let that be common knowledge?"

"Please do not feel slighted by this, but Forsythe was not necessarily pursuing you so much as your crown."

"Okay, I kind of get that. I realize it is not me but my power he is after. But did he start telling everyone that even before I arrived in Avonridge?"

"To my knowledge he never told anyone about his designs on the new queen, just his pursuit of the old one."

Marissa stopped walking and looked at Ruth, confusion and disbelief on her face. "Ruth, do you mean he tried to get Gertrude to marry him?" she asked, incredulous.

"He did indeed," she replied. "He had been trying for over ten years to woo Queen Gertrude. Of course, being quite mature, she saw him for what he was; a young upstart with designs on the throne."

"But, why did everyone get so upset when he started talking to me? Didn't you all know I would see right through him also?"

"We meant no disrespect, Majesty, we just did not know how you would react to him. We could not be sure if you would understand what he was after."

"Well, I understood almost immediately."

Ruth chuckled. "I remember. He was none too happy when he took your leave."

Marissa laughed at the memory as well. "Yes, I was actually a bit surprised when he laid his cards on the table so early," she said. "You would have thought he would have waited until we became friends or something."

"One would think," Ruth agreed. "However, Forsythe is not one for waiting. He is an impertinent man, and that must be why he was so quick to approach you."

"So, he actually tried to get Gertrude to marry him," Marissa said. "I must admit, I am impressed with his nerve. You have to hand it to him, he's got brass."

"Brass?" Ruth said. "What is brass?"

"Oh, just another word for nerve, guts, that kind of thing."

"Well, you are correct, he does have brass. He was a constant thorn in her side for quite some time. She would even get so annoyed with him at times that she would order him out of her sight."

Marissa laughed. "That is a good idea," she said. "Did it work?"

"Only for a while. A few weeks would go by and he would be right back again. At every banquet or party, he would ask her to dance. Sometimes she even accepted. After all, he is a good dancer. And no one else in the kingdom would ask her. The only other time she could dance was when royalty from neighboring kingdoms would visit. And Queen Gertrude loved to dance."

"Well," Marissa said, "a thought just occurred to me. Since Forsythe professes such a desire to be married, perhaps I should arrange for it to happen."

"Majesty," Ruth gasped. "You cannot mean such a thing!"

"Oh, relax. I'm not suggesting I be the one to marry him. No, of course not. But I have noticed the way Bethany looks at him when she thinks no one sees. Perhaps she would be agreeable to a royally decreed marriage."

Ruth nodded. "Yes, that is a marvelous idea. In fact, it should prove quite interesting to watch unfold." She chuckled. "I am quite eager to see the look on Forsythe's face when he is told."

Marissa chuckled as well. "Yes, I like the idea. I believe I will speak with Bethany this evening, and if she is indeed agreeable, I will call Forsythe to the palace and inform him of his royal duty."

FORTY-ONE

Bethany entered Marissa's chambers with obvious trepidation. She felt she was still under the gun for having fallen asleep while tending Kevin, and she was afraid she'd be punished. It had been several weeks since the incident, and Marissa had only scolded her, but every time she was in the queen's company, she was afraid.

Walking into Marissa's chambers, Bethany kept her eyes cast downward, and upon entering, dropped to a low curtsey and remained there for several moments, looking at the floor in fear.

Marissa chuckled to herself, trying not to let the nervous lady-in-waiting see her amusement. She realized that Bethany was waiting for her to tell her to rise, and quickly did so. When the young woman was upright, Marissa began.

"So, Bethany," she said amiably, "how have Kevin's lessons been going?"

"Oh, ah," she stammered nervously. "I've been doing my best. I have not fallen asleep once since that time."

"I am quite sure you have been doing a good job," Marissa agreed. "In fact, I am sure all of you ladies have been doing quite well."

Bethany was silent, waiting for Marissa to continue. Her face had gone pale and she seemed to Marissa to be trembling slightly. The queen finally took pity on the girl and smiled.

"Bethany, I did not send for you to scold you. In fact, I am hoping what I have to say will be most welcome."

Bethany's color came back somewhat with the relief she felt, but she was still quite anxious.

Marissa continued. "I have sometimes watched you when you happen to be in the company of Forsythe. I have reason to believe there are feelings there. Am I correct in this assumption, or am I mistaken?"

"Oh, Majesty," she blurted out, "I would never disregard my duties just because a man is present. I assure you my feelings would never get in the way of my teaching young Kevin."

Marissa was slightly taken aback by Bethany's misunderstanding. She put up a hand to silence the young woman. "Bethany," she said gently. "I did not mean to imply anything like that, I was simply stating an observation. Is my observation correct?"

Bethany blushed deeply and cast her eyes down again. Her mouth moved, but no words came out.

"I assure you," Marissa said, "I'm not in the least upset by those feelings. I only wish to know, so I may make plans that I have been considering."

Bethany looked up, curiosity in her eyes. "Yes, Majesty," she said sheepishly. "I have very strong feelings for Forsythe. Unfortunately, I do not believe he returns them."

Marissa nodded and thoughtfully paced up and down. "I see. But your feelings are deep?"

"Yes, Majesty."

"So, Bethany, how would you feel if it were a royal decree that Forsythe take your hand in marriage?"

The girl gasped and put her hand to her lips in surprise. "Do you mean that, Majesty?' she whispered through her fingers. "Would you truly decree that he take me as his wife?"

"I've been toying with the idea," Marissa said, still pacing as if continuing to weigh the pros and cons. "So that would be a welcome duty to you?"

Bethany bounced on the balls of her feet for a few seconds, then stopped and composed herself. She cleared her throat and said, "Yes, Majesty, that would be a most welcome duty indeed." Then her face darkened with concern. "But Majesty, what if Forsythe does not welcome the duty?"

Marissa stopped pacing and faced the girl again. "That is a very real possibility," she agreed. "That is something you must consider. I don't know how he will take the order, and he might turn his anger toward you. Rest assured I will not hesitate to step in and hold him accountable for his actions, but it could make for some very uncomfortable living arrangements."

Bethany smiled. "Oh, it may be difficult at first, but I am sure I would be able to help him accept his situation. I can be fairly persuasive when I want to be."

"That is good to hear," Marissa said, smiling. "I will arrange it."

Marissa would not even consider meeting with Forsythe in her chambers. She sent for him and waited in the banquet hall, sitting upon her throne. Merlin sat in her lap, head held erect, looking as if he were posing for a portrait. Besides the little cat, there was no one else in the hall. She wanted no one to witness this encounter, as she had no way of knowing just how Forsythe would react.

She wore her gleaming white gown with the jeweled bodice, and had her purple cape draped over her shoulders. Her rich brown hair was hanging in lustrous tresses down her back, and the gold and purple crown sat atop her head. She was purposely in full regal splendor, allowing for no mistake as to her authority.

Forsythe walked up to her and bowed. She silently noted that his manner was not arrogant or mocking, and she wondered if he realized his error of previous encounters with the queen. She quickly pushed that thought aside however, when she remembered his all-consuming self-promotion.

"Ah, Majesty," he said, "you look absolutely breathtaking." And he again handed her a rose.

She accepted it with a half-smile. Yes, she thought, this would prove to be an interesting encounter. As she grasped the rose, Merlin gave an almost imperceptible growl, obviously no happier in Forsythe's company than she. She stroked his glossy fur as she sat back, twirling the rose between her fingers. She didn't bother to thank Forsythe.

"So," Forsythe said. "I was summoned. You must have need of me."

She took her time answering. After several seconds, she softly said, "Yes, I have need of you."

Forsythe's countenance broke into a big smile, and his yellow eyes twinkled. "I was sure you would see the advantage of accepting my offer, dear lady." He gushed. "It was only a matter of time."

"Do not jump the gun," she cautioned lightly. "I am about to accept your offer, but it may not be exactly what you have in mind."

"Details can always be worked out at a later date," he said confidently. "The important thing is, my dreams are about to come true."

"Yes, you have been dreaming of a marriage for quite some time, have you not?"

"Indeed, I have, my dear," he answered. "And you should be as well. Marriage is a joyous union and can only benefit both participants."

"You don't know how happy it makes me to hear you say that," she said, smiling with an almost conspiratorial gleam.

Forsythe smiled even more broadly, and dipped his head in acknowledgment. "It is a fact. When two people are joined in holy matrimony, they are bound together as if by one heart, one mind. So, when shall we make the announcement?" he asked.

"Oh, we can make the announcement this afternoon," she said. "That is, if Bethany is agreeable to the timing."

Forsythe's smile faltered slightly and his eyes clouded over with confusion. "Bethany? Your lady-in-waiting? What could she possibly have to say about it?"

"Well, she should most definitely be a part of your wedding announcement."

Silence filled the air for several moments. "I am not understanding, dear lady," he said hesitantly. "You referred to it as my wedding announcement, not ours."

"Oh, it is not our wedding announcement of which we speak," she said easily. "As I have told you before, the queen of Avonridge does not marry. No, this will be the announcement of your marriage to Bethany."

Color rushed to Forsythe's face as anger burned hot in his eyes. "Bethany? What trickery is this?" he demanded, standing and stepping back a pace.

"Trickery, Forsythe? There is no trickery here," Marissa cooed sweetly.

"You led me to believe you were accepting my marriage proposal, and now you offer Bethany in your stead. This most certainly is trickery!"

"I led you nowhere, Forsythe," she calmly countered. "If you jumped to that conclusion, I cannot be blamed. In fact, I informed you at the beginning of our conversation that this was not exactly what you were expecting. If you chose to ignore that statement, I cannot be held responsible for your mistake."

He backed up two steps and angrily stated, "I will not marry that simple-minded cow!"

At that moment Marissa was both relieved that Bethany was not present, and angry at Forsythe's insult. Even Merlin seemed to take exception to his words, for the little cat hissed loudly and gave a high-pitched snarl.

"Forsythe," Marissa said loudly, "you do not seem to realize, I have not asked you." He paused in his anger and looked at her for a moment. "I have not asked you," she said again. "I have told you."

He jerked his head back as if he'd been slapped, but said nothing.

"This is not a request," she continued. "This is a royal decree. You will do your royal duty and take Bethany as your wife. And you will do so in a manner befitting your noble birth."

"My noble birth?" he spat. "Whatever could you possibly mean by that? What does my noble birth have to do with any of this?"

"None of this is Bethany's doing," she said, leaning forward slightly. "I am the one who thought of this alternative, and I made the decision. So therefore, the anger you feel now, and whatever anger you may feel in the future, will not, I repeat, will not be turned toward that young woman. You claim highborn status, show me it is so. You will treat her with respect, dignity, and even tenderness."

He stared in disbelief at her, his eyes blazing and his mouth drawn so tight his lips were thin white lines. She sat calmly, allowing the situation to sink in. After a while, he took a deep breath.

"I came here today with the belief that I would be marrying the woman I love," he said. "But now you order me to marry a witless fool instead. I do not believe the royal influence should be allowed into such a personal and intimate area of one's life. There should be limits, even for the crown."

"Personal and intimate area of one's life," Marissa echoed. "Would those limits even reach to the High King?"

Confusion reflected in his eyes again. "Why, yes," he said hesitantly. "Even the High King should allow matters of the heart to be personal."

"Is that what you told King Arthur while you were asking him to order me to marry you?"

He shook his head, and some of the anger turned to anxiety. "He told you what I requested?"

"Oh yes, he told me," she said lightly. "Why does that surprise you?"

He now feigned anger as he said, "The conversation was between the High King and myself. He had no right to relate to you what we discussed privately."

"Indeed," she said, smiling. "How hypocritical."

His feigned anger again turned real and he said, "I have had enough of this conversation. With your permission, I wish to take your leave." And he turned to go, but seemed to think of something, and stopped himself. He threw his arm out to the side and was preparing to perform the same exaggerated bow with which he had insulted Marissa on their previous encounter, when she held up her hand to stop him.

"Two things before you go, Forsythe. First, remember my words about not turning your anger toward your new bride, and second, I would strongly recommend that you," she lowered her voice to a hard-edged warning, "remember your place."

The hand that was held out, he slowly lowered to his side. His eyes showed unmistakable understanding at her words, and he performed a respectful bow. He then took several steps backward to an acceptable distance before he turned his back on his queen, and quickly left the banquet hall. As his scarlet waistcoat disappeared, Marissa was again stabbed with a nagging feeling, but she again chalked it up to the encounter. Meeting with Forsythe was beginning to tax her nerves. She would be glad when it would become a rarity.

FORTY-TWO

Bethany was ecstatic at the news. She bounced up and down on the balls of her feet with her hands clasped to her breast.

"Oh, Majesty," she said breathlessly, "this is such a wonderful day. I cannot begin to tell you how happy you have made me."

"Be careful, Bethany," Marissa warned. "Forsythe was not at all pleased."

The knowledge that the man she would be marrying didn't love her only made her face fall for a moment, and then she was smiling broadly again.

"It does not matter," she said. "As you said, it may be difficult at first, but he will come to see the good in it."

"I certainly hope you are correct in this," Marissa said truthfully. "I would hate to think I had condemned you to a life of misery."

"Oh, it will not be miserable," Bethany assured her. "I will be able to persevere and win his affection, if not completely his love."

Marissa smiled and nodded. She found Bethany's words to be mature and insightful. She may never be loved by Forsythe, and she was completely aware of that. She had thought Bethany was childish and spoiled, but she gained a new respect for this young woman that was not wearing blinders. She knew it would be difficult, but was willing to try.

Still, Marissa was a little worried for the girl. Forsythe did not appear to be the kind of person who cared about others. She'd only seen him be selfish and manipulative. She hoped Bethany would do as she'd promised and let Marissa know if he was ever to become abusive.

The banquet hall was again adorned with lively colored drapes hanging over the stone walls. A stringed quartet played beautiful music fit for dancing, and the guests were enjoying themselves completely.

Bethany shone in an enchanting, soft blue gown that amplified the blue in her eyes. Her dark hair was piled on top of her head with soft curls falling down to frame her smiling face. She was in her glory.

Forsythe, on the other hand, was standing against the wall with a scowl on his face, his arms crossed over his chest. His eyes were hooded and cast downward, as if he were awaiting his execution. He did have the good sense, however, to make polite conversation whenever someone would stop to chat.

Marissa wasn't looking forward to it, but she made her way to the sullen groom-to-be. When she stepped up to him, he pushed himself away from the wall and stood ramrod straight, looking down his nose at her.

"No need to be quiet so formal, Forsythe," she said sweetly. "After all, this is your engagement party. You should be relaxed and having a good time."

"I do not think it appropriate to be relaxed in the presence of my queen," he said coldly. "After all, I do need to remember my place." The edge he put on his words, the blade of a knife.

Marissa laughed easily, refusing to be baited, and said, "Of course, of course. But then again, everyone needs to remember their place, do they not?"

"Apparently some more than others," he said.

She again ignored his remark. "I do enjoy having these parties," she said. "They are so much fun, don't you agree?"

He merely grunted.

"Bethany looks stunning, does she not?" Marissa said. "I do believe this arrangement has been quite good for her. She has grown up a bit since I gave her the news. I am sure you will be quite happy with her."

He made no comment.

"And please remember," she said, her voice turning hard, "I will not tolerate any retaliation against that young woman."

"You have made that quite clear, Majesty," he replied coldly. "Rest assured, I will not direct my anger toward the young half-wit. In fact, I will turn no emotion toward her."

"Ah, I see you are coming around," she said derisively. "I am sure your home will just be packed full of love and laughter." Then her voice became softer, and she said, "Keep in mind, Forsythe, life is mostly what you make it. If you choose to be miserable, you will be miserable. I would strongly recommend choosing happiness."

"Wise words from a woman who walked into the role of queen less than a year ago. I would expect life is quite happy for you when everything is handed to you."

"I will not argue how lucky I am," she said. "I was given a wonderful life just by chance. But yours is not so awful. You have land and money, and you will soon have a wife who loves you. She will do everything in her power to make you happy. It would truly be a shame if you do not let her." And she turned and walked away.

Walking through the crowd of revelers, many called greetings to their queen. She came upon the group of ladies-in-waiting, without Bethany, as she was being fussed over by some of the women of the realm, and she overheard the ladies as they chattered.

It seemed all she heard was griping and complaining, and she stopped for a moment to listen. She should have listened instead to her inner voice and kept walking, because the moment they spied her they were on her like a pack of wolves. Funny she should liken the ladies to those wolves.

"Majesty," Louisa was the first to nearly shout. "I too would love to have a husband. With whom can you match me for marriage?"

"I have been telling her," Jane, an older lady, said, "that she should count her blessings she does not have a husband. They are nothing but trouble."

"I would like to be married as well," Wanda cut in. "Jonathan's wife has been dead a year now and he needs someone to help him with his two babies."

"Since you are making matches," Angelina added, "I wish to marry also."

The ladies-in-waiting all surrounded Marissa, making her feel trapped.

"Ladies, ladies," Marissa said, holding up a hand for silence. "I have not become the kingdom matchmaker."

"But you made a match for Bethany," protested Louisa. "Why can you not do the same for me? After all, Forsythe is a man of wealth and class, a good husband for anyone, and you chose Bethany." The disdain was clearly evident in her voice.

"She simply will not listen to me," Jane said. "She should count herself lucky to not be encumbered with a husband, as they are nothing but misery."

"I still want one!" insisted Louisa.

"As do I," agreed Wanda. "Why can we not be married as well?"

"I do not know why you had to choose Bethany," Wanda continued. "She is addlebrained and childish. Why could you not have chosen me instead?"

"Please, please," Marissa cried feebly. "It was not a decision made lightly. I could see that Bethany had real affection for Forsythe. That is why I chose her."

"But I could have had real affection for him if I had known you would be choosing a wife for him," wailed Wanda. "Could you not have asked us first?"

"Ladies, really, I did not mean to cause such discord," Marissa said. "I only wanted to give Forsythe something to occupy his time."

"But I could occupy his time much better than Bethany could," Wanda countered. "After all, I am older and more worldly. I would have been a much wiser choice to be matched with Forsythe."

"But I thought you were all very happy being my ladies-in-waiting," Marissa said slyly. "Is that not enough of a job for you?"

"Exactly what I have been trying to make them understand," Jane agreed. "With their duties to Your Majesty, they would not have sufficient time for the burden of a husband."

"I could do both," Wanda said. "Many ladies-in-waiting have been married, and their duties have not suffered."

"I could manage both as well," Louisa said. "There would be no problem doing both."

"As could I," Angelina chimed in. "It would be no problem keeping a husband and being a lady-in-waiting. If I could choose whom I would like to be married to, I would choose Oliver. He is handsome, and his family is quite respected."

"And wealthy," Jane added. "Of course, you would choose a wealthy husband. Who would willingly choose to be poor? Besides, Oliver has been keeping company with Wilma, who is also from a wealthy family," she said snidely. "He is not in the market for a wife, he has already decided on one."

"But there are others," Angelina countered. "All we need to do is choose an acceptable match and I may be married."

Marissa's head was beginning to spin. She could understand the ladies' wishes to be given the same special treatment as one of their peers, but she'd had no idea of the hornet's nest she would rattle by her decision to disencumber herself of Forsythe. She hadn't really thought about it, but if the ladies were of a mind to marry, wouldn't they have already been looking?

Of course, Louisa had made known her desire to marry the High King Arthur, but if everyone would be reasonable, they would realize what a long-shot that was. But now, with the ladies all around her, pecking at her like a flock of hens, she was almost regretting the decision to marry Forsythe off. She had to think of something to get them off her case.

"Ladies," she tried again. "I understand your feelings. Believe me, if I were in your shoes, I would probably want the same thing. Unfortunately, I do not want to go around forcing the men of the realm to take wives. It came as quite a surprise to Forsythe when I informed him of my decision, and I am in no hurry to do that again."

"Do you mean he was not happy with your decision?" Wanda asked hopefully. "I am sure he would be much happier if you were to choose me instead of Bethany."

"If that is the case," Angelina cut in, "why not choose me? He would be happier with me, not Wanda."

"I believe I should be the one chosen," Louisa said. "I have much more in common with Forsythe than either of you."

"If you want to go by who has more in common, the choice would clearly be me!" Wanda insisted.

Marissa was beginning to feel as if she were drowning, when Ruth's sharp voice cut in.

"That is quite enough, ladies! Do not badger Her Majesty with your petty griping."

Marissa turned a grateful eye to her rescuer.

"Her Majesty does not need to explain her decisions to you," Ruth said. "She has her reasons, and you will abide by them."

"But Ruth, we only want what Bethany has been given," Louisa protested. "It is only fair."

"Fair has no bearing here," Ruth said harshly. "The decision has been made and there is no cause for debate. Now, leave Her Majesty in peace and get back to your inane sniping. Honestly, you are all a bunch of spoiled children."

Wanda began to protest, but Ruth held up a hand to silence her.

"I said, 'go'," she said, and they unwillingly turned to leave, each muttering softly about the unfairness of it all, except of course for Jane, who had the look of someone who thinks they know everything, but no one will listen.

"Oh, thank you," breathed Marissa. "I was starting to think they were going to tear me to ribbons."

"You should not be so gentle with them, Majesty," Ruth advised. "They need a strong hand or they will walk all over you."

"Yes, I'm beginning to realize that," she agreed.

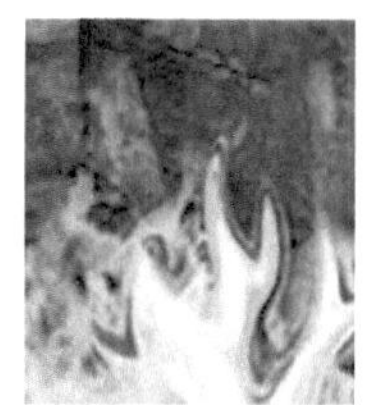

FORTY-THREE

Marissa and Ruth walked through the banquet hall, occasionally calling greetings to the guests of the party, but mostly enjoying the peace after the ladies were dispersed. Marissa chuckled to herself. This idea of hers had certainly caused a tumult. It seemed the only ones happy about it were Bethany and herself. Oh well, that was the advantage of being queen; you could think about yourself for a bit and everyone had to obey.

She really didn't think Forsythe was in such a bad situation. As she had told him, he would soon have a wife who loved him. She was sure Bethany could win him over, at least she hoped.

They continued to enjoy the party. The music was festive and several couples were dancing. Suddenly, Marissa was stricken with such a feeling of dread that her knees almost buckled. Ruth saw her stagger, and quickly grabbed her arm to support her.

"Majesty," she gasped. "Are you unwell?"

Marissa stood for a few moments, trying to catch her breath, leaning on Ruth's supportive arm.

"Majesty," Ruth said again. "What is it?"

"Something is wrong," she said. "Something is happening that should not be."

"You do not mean the marriage, do you, Majesty?" Ruth asked, confused.

Marissa shook her head. "No," she said. "Not that. But something is happening somewhere. Where is Kevin?"

"In his chambers with Amber," Ruth said. "Right where you said he should be."

Marissa shook her head again. "I must see him," she said, and before Ruth could respond, she started walking toward the doorway that led to the royal chambers, her feet barely touching the stone floor in her haste. As she hurried through the crowd, she drew concerned stares from many people, but she was moving too fast for anyone to speak to her.

Reaching the first corridor to the royal chambers, she dashed through the doorway, the skirt of her luminous white gown billowing behind her like a sail. Ruth was close on her heels, and almost running to keep up, concern etched on her face. She was glad the corridor was deserted; she knew the queen would not stop to speak to anyone.

"Majesty," she called. "I do not understand the rush. What could possibly be wrong?"

"I don't know," Marissa replied. "But I need to see Kevin, need to see that he is all right," and she continued down the corridor. Just as she reached a corner and was prepared to turn into the next corridor, she stopped so abruptly that Ruth bumped into her from behind.

"Forgive me, Majesty," Ruth gushed. "I did not know you were stopping."

Marissa seemed not to hear, and asked, "Do you smell smoke?"

Ruth sniffed the air, but before she could reply, Marissa was off at a run, dashing down the corridor towards Kevin's chambers.

Ruth hurried after her, trying but unable to keep up. Marissa ran with all her strength, and as she ran, she began to see the smoke she had smelled. Rounding the last corner that led to Kevin's chambers, she stopped in horror at the sight of flames leaping out the door.

"Kevin!" she screamed, and resumed her flight with Ruth behind, calling after her.

"No, Majesty, you must not!" But Marissa paid no heed.

Standing in the corridor past the door to Kevin's chambers was Amber, the lady-in-waiting that was tending Kevin for the evening. She was in an obvious panic, wringing her hands, stamping her feet and crying hysterically. When she spied Marissa, she screeched, "Oh, Majesty, help him!" And she dashed back and forth from wall to wall at a frenzied pace. "Help him, please," she wailed. "Help the child. Help him!"

Marissa ran to the door, standing on the opposite side of the corridor to avoid the flames. She tried to look into the chamber, but the flames blocked her view. Taking a cautious step closer, the heat began to sear her face. Finally, she spied the little boy. He was standing in the center of the room, staring at the flames that danced all around him, but did not touch him. His hands were clenched into tiny fists and a look of abject terror marred his cherubic face.

Marissa wanted to scream, wanted to run through the flames to snatch him away to safety, but knew her gown would ignite once it came into contact with the fire. Fleetingly she wondered how the flames could be all around Kevin, but not actually touch him. Then she wondered how the flames could continue on the stone floor and walls. They appeared to be dancing over the stone tiles, continuing to burn with no apparent fuel.

Knowing she had only seconds to save her precious boy, she turned to Amber and ordered, "Get the bucket of water from my chambers." When Amber hesitated, she yelled, "Now, Amber!" and the young girl spun on her heel and dashed down the corridor to Marissa's chambers.

She then turned to Ruth and commanded, "Go to the sitting room around the corner and bring me the blanket lying on the lounge."

Ruth was off in a flash to obey.

Marissa turned back to the room, her heart aching with the frustration of not being able to just rush in and scoop him up in her arms.

Kevin was pumping his fists up and down, as if beating an invisible object to the ground. She was about to call to him to reassure and comfort him, when she noticed a difference in the flames. Each time he punched his fist downward, the flames edged away from him by several inches. He punched downward four or five times, and the flames receded by more than a foot. When he paused, the flames crept back toward him. Punching downward again, the flames ebbed away; pausing, they resumed their march. She watched in awe as he seemed to exert some power over the flames, but could not quite master them.

Amber and Ruth arrived at the same moment, and Marissa quickly took the blanket. Amber put the bucket on the corridor floor, and Marissa was pleased to see that she hadn't spilled much at all. Pushing the woolen blanket into the bucket, it thirstily sucked up the water, becoming drenched. When all the water was absorbed, she tossed the blanket around herself, careful to get as much of her gown under the sodden material as she could.

Ruth realized what Marissa was planning, and begged, "Majesty, you must not. It is too dangerous."

"Ruth," she replied, "how can I not?"

Amber stood off to the side, again crying and wringing her hands, and Ruth grasped her wrist and practically threw her down the corridor toward the main part of the palace. "Go, get help," she ordered, and to her credit, Amber took off at a run.

"Majesty," Ruth tried again. "We cannot do without you, you are most important to the kingdom."

"Ruth, I couldn't live with myself if I didn't," and she moved toward the door to Kevin's chambers.

The fire seemed to possess a consciousness, and it surged up and leapt toward her. She pulled the blanket tight around her head, exposing only her eyes to see where she was going. Flaming fingers reached toward her, as if trying to rip the protective blanket from around her. She held tight and pushed forward.

Kevin, still thrusting the flames down around him, looked up and spied Marissa. "Mama!" he screeched in terror. "Mama!"

"I'm coming!" Marissa called. She continued into the wall of fire, and behind her she could hear Ruth's anguished cries.

"Oh, Majesty, you should not do this!"

Ignoring the woman's protests, she began moving faster. The flames danced around her, completely engulfing her. A moment of panic caused her to falter, but she heard Kevin's cries, and resumed her mission. The heat of the fire penetrated the sodden blanket, and she wondered how long it would take for the material to dry and become flammable.

Knowing that was, indeed, a real danger, she practically ran to the little boy who stood terrified amid the flames. As she hurried across the floor, she vaguely realized the stone tiles that were aflame were not singeing her slippered feet. In fact, that was the only part of her that felt untouched by the fire.

Finally breaking through the wall of flames, she stood next to the boy who held his arms up, frantically begging to be picked up. She opened the blanket and scooped him up in her embrace, completely covering him with the wet wool.

The fire seemed to grow angry, and its intensity doubled, sending flames leaping all around the two. Marissa began to feel lightheaded, and knew she had to hurry. Turning toward the door, she again felt panic surge within her as she couldn't see the door through the flames. She became disoriented and froze in place, not knowing exactly which direction to take.

"Ruth," she screamed. "I'm lost."

"Majesty, I cannot see you. You must hurry out."

Hearing Ruth's voice greatly calmed Marissa, and she took several steps toward it. "Ruth," she called again. "Keep talking. Your voice will guide me."

Ruth quickly obeyed, and called, "I am this way, Majesty. Keep moving toward my voice."

Marissa walked faster, emboldened by the assistance she was receiving.

"Hurry, Majesty," Ruth said. "Hurry toward my voice. The flames stay only at the doorway. Once you are through, you will be safe."

Finally, Marissa could see the open doorway and ran through it, just as the heat and smoke overcame her. Her knees buckled and she pitched forward, throwing down a hand to try to break her fall, while holding tight to Kevin, desperately trying to avoid dropping or crushing him. She felt her hand slam onto the stone floor, jolting her shoulder, and then all went black.

Marissa awoke with a start. The first thing she noticed was her throat felt dry and sore. The room was dimly lit, making her wonder exactly where she was. It took several moments for her to recognize her own bedchamber.

Kevin was sitting next to her on the pillow beside her head, and when she looked at him, he broke into a big grin and happily shouted, "Mama!"

Marissa sat up and hugged the little boy to her chest, smothering his head with kisses.

"Majesty." She recognized Ruth's anxious voice. "Majesty, are you well?"

Marissa looked at her most trusted friend and managed a weak smile.

"Yes, Ruth," she said in a dry, raspy voice. "I'm just fine." She still cradled little Kevin to her chest, and he became antsy and, laughing, pushed away from her.

Looking past Ruth to the open doorway of her bed chamber, Marissa saw several of her ladies-in-waiting crowded around. Standing behind the anxious-looking women, she spied the leather-clad head of the Black Knight. Though only his eyes were visible, even at a distance she could see he was greatly distressed. When their eyes met, she thankfully saw his visage soften. She could see the relief in his eyes when he realized she was indeed fine.

When the ladies saw that Marissa was sitting up, they began clucking like the barnyard chickens Marissa thought them to be.

"Oh, Majesty," clucked Jane, the older lady. "We were so worried, we did not know if you would be all right." And Marissa saw in her eyes true concern.

"Yes, Majesty," Louisa added. "We are so relieved you seem to be well."

"Oh yes, Majesty," Wanda echoed. "Thank the Lord you did not get hurt. We would not know what to do without you."

Marissa believed that statement to be true, since there was no provision for an heir to the throne. She wondered just what would happen if she met with an untimely death.

"Well, as you can see, Her Majesty is fine," Ruth said. "Now all of you leave her chambers so she can rest after her ordeal."

The ladies-in-waiting all began to protest. "But we wish to be with Her Majesty to be sure she is well," Louisa complained.

"As I said," Ruth insisted, "you can see she is well. However, she needs her rest. Now go, all of you."

Fearing any repercussions for not obeying, the ladies turned to go, grumbling all the while about the unfairness of it all. When they had dispersed, only the Black Knight still stood at the doorway. He remained for several moments, then dipped his head in respect, and turned on his heel and strode away.

Once he was out of sight, Ruth turned back to Marissa and gushed, "Majesty, it is truly a miracle you were not hurt, or even killed. The fire was like a living beast, trying to devour you and little Kevin. I was sure we would lose you. I was afraid for your life and the safety of our kingdom. What would we do without you?"

Then her voice turned scolding, like that of a mother with a rambunctious child, and she said, "Avonridge needs you, Majesty. Our kingdom would be thrown into turmoil if anything were to happen to you," and she accentuated her words with a wag of her finger.

"Yes, Ruth," she said gently. "I thought about that myself. I would be putting Avonridge in a very bad position if I allowed myself to be harmed. However, and you know this is true, I could not let anything happen to Kevin."

Ruth nodded. "Yes, I understand. And please do not think me coarse or cruel, but you are infinitely more important than the child, no matter our feelings for him."

Marissa noted Ruth had said 'our feelings', not 'your feelings,' and she did appreciate that. She also understood what Ruth was saying about her importance. As painful as it would be to lose Kevin, it would not put Avonridge at risk. The loss of the queen certainly would.

Ruth began fluffing Marissa's pillows and fussing over her and Kevin. All the while, the little boy sat quietly, grinning like he'd just won a prize. Marissa wondered if he might understand the gravity of what had occurred, and knew just how lucky he was to be alive and well. Children were very intuitive.

Finally, Marissa had had enough fussing, and said, "Ruth, I need to see Marcellus. Please send for him."

"At this hour, Majesty?" she protested, clearly not relishing the idea of being anywhere near the magical old man. "It is near to midnight. You were asleep for well more than an hour."

"Near to midnight," Marissa repeated. "I suppose it can wait until morning."

FORTY-FOUR

Marissa paced while waiting for Marcellus. Kevin sat on a chair at the table, happily playing with the wooden animal figures he had received for Christmas, and singing to himself. Every so often she would look at the little boy and feel a maternal pang so strong her breath would catch in her throat. The thought that she had come so close to losing him made her heart ache. To think that he was almost ripped from her life was simply unbearable.

While Marissa was standing still, just gazing at Kevin, the little tyke looked up at her and smiled, his whole cherubic face brightening in an angelic glow.

"Mama," he said happily. Realizing he seemed to have no lasting ill-effects of the fire, relief flooded Marissa's heart. She hoped he wouldn't be encumbered by nightmares of the incident.

She resumed pacing, wondering aloud where the old magician was and why he was taking his sweet time in obeying her summons. Kevin seemed to think she was talking to him.

"Mama, Marcellus," he said.

She smiled at him, and said, "Yes darling, I'm waiting for Marcellus."

As if on cue, a stooped figure hobbled through the open door. "You sent for me, Majesty?" he said. His voice still holding a note of scorn, however slight.

Marissa stopped pacing and stood facing the old man. She did not answer right away, deciding to make him wait as he had made her wait. If Marcellus didn't know about the fire, when he found out about it, he would use it as an opportunity to try to take Kevin to live with him. Marissa couldn't allow that, couldn't bear it.

Marcellus waited patiently for Marissa to inform him of why she had summoned him. Finally, she obliged.

"We've had an incident here," she said.

"Yes," he replied. "A fire."

She nodded. "So, you've heard."

"I hear most things that go on at the court of Avonridge," he said matter-of-factly, as if it was simple common knowledge.

"I'm sure you do," Marissa agreed. "However, I have reason to believe you may be able to shed some light on some of the mysteries surrounding the fire."

"Me?" he asked, feigning incredulity. "What information could I possibly contribute that would be of any use?"

They continued to stand facing each other; Marissa refused to offer this man a seat.

She began to pace again, and said, "There was an element to the fire, in fact several elements to the fire, that caused me to believe you could answer some questions concerning it."

"Of course, I will be happy to assist in any way I can," he offered, and Marissa caught only a hint of, what, sarcasm, defiance?

Deciding to ignore the emotion he had put into his offer and accept it at face value, she said, "Good. I want you to tell me how the fire survived on the stone walls and floor."

Marcellus seemed somewhat surprised. "On the stone?" he repeated.

"Yes," she answered, pacing back and forth before the old man. "The fire blazed without any obvious fuel. How?"

He shook his head. "I did not receive all the facts," he softly whispered, as if to himself. "When I heard of the fire," he said, to the queen, "I assumed it fed on curtains or furniture. I did not know it blazed on air alone."

Marissa stopped pacing and looked at the magician. She felt in her gut that he was sincere, and believed he was hearing this information for the first time.

"Then you didn't know Kevin pushed it back?"

"Pushed it back?" he repeated, somewhat confused. "What do you mean 'pushed it back'?"

Marissa related what she had seen when she looked into the room, the fire surrounding Kevin on the stone tile floor, and the small circle of bare floor where the fire did not encroach. She told Marcellus of how Kevin could force the fire back by making pushing movements with his tiny fists. All the while Marcellus listened, enrapt by what he was hearing.

When Marissa finished speaking, a look of amazed wonder settled on the old man's face.

"Remarkable," he breathed. "This is truly remarkable. I did not know this, Majesty, but it truly is joyous news." His voice held the same wonder and awe that was still upon his wrinkled old countenance.

"Explain this to me, Marcellus," Marissa said. "Why is it so surprising, given Kevin's talents at levitation?"

"Ah, levitation. That is simply controlling inanimate objects. There is not a will to be dominated, the objects have no will. But fire, fire is a living, breathing thing."

"Living and breathing?" she echoed. "Fire is not alive."

"Ah, but it is, Majesty," he insisted. "It feeds, it grows, it thinks. Fire has a will all its own. It is very selective in what it devours."

Marissa wanted to argue, but if she opened her mind, she could actually believe that fire was a living thing. It seemed to be a creature filled with rage. At least, it seemed such when it had encompassed Kevin in his chambers.

Marcellus' face became dark and concerned. "You said the fire surrounded Kevin," he said.

"Yes, and it seemed to want to get at him. He pushed it back, only to have it rush toward him again."

"It wants to finish what it started in his parents' cabin," he said softly, shaking his head in dismay.

Marissa remembered that Kevin's entire family had perished when fire burned their cabin down. Could the old man be right? Could fire have such an awareness that it would want to finish a job?

"But why would fire want to destroy an entire family?" she asked anxiously. "What could possibly be the motivation for such a heinous act?"

Marcellus was pensive. "There is only one explanation that comes to mind," he said darkly. "There is another wizard within your realm."

"But wouldn't everybody know of another wizard?" she demanded. "If not everyone, at least you and some members of my household. Perhaps Ruth or Sir Erick? They seem to know everything around here. Besides, why is another wizard a problem?"

"Another wizard, dear lady, would not welcome the competition. And if his powers are significant, he can go unnoticed from the general populace. He must have seen the talent in young Kevin, and decided to eliminate him."

Marissa looked at the angelic little boy that was making his toy horse gallop across the table, lost in the pleasure of his pretending, and her heart tightened.

"Someone wants him dead," she whispered, dread making her voice a hoarse croak.

"Indeed," Marcellus answered, just as anxiously concerned. "And he may not be an all-powerful wizard, but he is formidable enough to be dangerous."

"So, how would we go about finding this wizard?" Marissa asked. "Find him and eradicate him."

"Yes," Marcellus said thoughtfully. "We certainly need to eradicate him. Majesty, the responsibility of removing this threat to your kingdom is completely mine. I shall take care of eliminating this menace. You need not worry yourself, I shall see to it."

"Fair enough," Marissa said. "Fight fire with fire, so to speak."

Marcellus nodded, a slight smile on his old lips.

"In the meantime," he said, "little Kevin must certainly come with me."

Marissa immediately became alarmed. She had been expecting this, but would not allow it.

"Why should he go with you?" she demanded.

"I can look after him much more efficiently. Here he is surrounded by addle-brained barnyard chickens. No one would know what to do in case of another incident."

"Well, you can forget that idea right now," she said. "Kevin stays here, he will be much safer here."

"I assure you dear lady, he will be safer within my care. I have the necessary skills to defend him from anything that may arise."

"Keep in mind," she countered, "I took care of the situation last night. I was able to get him away from the fire and keep him safe. You will be otherwise occupied trying to flush out and get rid of another wizard, a wizard that wants Kevin dead. How do you plan to keep Kevin safe while possibly doing battle? No, he stays here with me where I can keep an eye on him."

"I understand, Majesty. However, when he is out of your company, he will be tended to by incompetents. I will be able to shelter him even while I am 'doing battle'."

"No," she said, shaking her head. "If necessary, I will keep him with me every minute, but he does not leave the protection of the palace."

"As you wish, Majesty," Marcellus said, bowing his head in acquiescence, and disappointment. "I will, however, be able to continue my tutelage of the boy?"

"Absolutely," Marissa quickly agreed. "He needs all the schooling he can get. If he is going to have someone sending fire to kill him, he needs to be properly trained to defend himself."

"Very good, Majesty," he said, obviously relieved. "Shall I commence today's lesson?"

Nodding, Marissa said, "Yes, that'll be fine." And she walked out the door, leaving the old magician to teach the young magician.

FORTY-FIVE

"Camille, I have need of you," Marissa called from the edge of the clearing. Merlin stood at her feet, head held high and tail standing erect with a curl at the tip.

Marissa waited for the old seer to emerge from behind the pelt door of her hut. After only a moment, the fur was pulled back and the thin figure appeared.

A broad smile split the old woman's wrinkled face when she spied Marissa. "Majesty," she called. "How good it is to see you. Please, come and sit down," and she hobbled to the bench at the side of the clearing.

Marissa and Merlin walked to the bench. Marissa sat down and Merlin jumped up to perch beside her. Camille sat as well, and rubbed Merlin behind his ear.

"So, my friend," she said to the little cat, "can you only visit me when our queen accompanies you?"

Merlin meowed and rubbed his head against Camille's shoulder in a gesture of apology. He looked duly chastised.

Camille then turned to face Marissa. "So, Majesty, what brings you to my humble home?"

"Well," Marissa began, "it seems there is some competition for the position of court wizard. I have reason to believe there is another within my kingdom who has powers."

Camille nodded thoughtfully. "That would be Silas," she said.

"You know of him?" Marissa asked anxiously. "Why didn't you ever tell me about him?"

"I saw no need to," she replied. "Silas lives out in the forests. He never seemed to want to be a part of the daily life of Avonridge."

"Well, he seems to want something," Marissa said. "I believe he sent fire to my palace to kill Kevin."

The old seer's eyes widened in shocked surprise. "I had heard of the fire," she said. "But I did not know it had been sent. How could this happen without my knowledge?" this last more to herself than to Marissa. Nodding, she continued. "Yes, I had heard of the fire, and how you saved the little one. Very brave, Majesty. I would expect nothing less."

"Thank you," Marissa said distractedly. "But tell me about Silas."

"Yes, Silas. He is not so old as Marcellus, but he has been a part of Avonridge for many years. His power was never very great, so I did not think him much of a threat. But if he has learned to summon fire, there is no telling what other abilities he has acquired."

"But why did you not know about him?" Marissa asked. "I thought you knew everything that goes on in Avonridge."

"Those with powers may keep their lives cloaked from me," she explained. "Until now I believed Silas was not worthy of my attention. I had felt nothing out of the ordinary regarding him. He obviously knew enough to shield himself from me."

"So, could he be a genuine threat?" Marissa asked. "Should I take precautions?"

"If he has targeted little Kevin, you need to be quite vigilant in protecting him. Silas must have a reason for wanting him gone."

"Marcellus seems to think Silas regards Kevin as competition."

"This could be true," Camille agreed. "Little Kevin has shown some talent. If Silas is wanting to ingratiate himself to you, he would indeed see the boy as an impediment to that end."

"But why have I not heard of him before this? If he wanted to get to me, either to be a part of my court, or harm me in some way, why has he not made himself known before this?"

"That is something only Silas can know," Camille said. "The whys of doing things in this world are many and varied. Only he who holds it in his heart can say what is there."

Marissa nodded, but was still frustrated with that answer. "So, in the meantime, how can I protect Kevin if this man can send fire to try to kill him?"

"Be very vigilant, Majesty," Camille said. "Never let him out of your sight. The women that watch over him are inadequate. You may wish to keep him always at your side until you are sure he is safe."

"Yes," Marissa agreed. "Even now I am becoming more and more anxious for him. He's with Marcellus, and he will protect Kevin, but I still worry."

"Then go to him and keep him safe. And, Majesty," she said softly, "beware, treachery."

"But I thought that's what we have been talking about."

"No, there is someone in your court who cannot be trusted."

"Tell me who, Camille," Marissa said. "Whom can I not trust?"

Instead of answering, the old seer stood up, petted Merlin on the head and said, "I'll be expecting a visit from you soon, my little friend." Then she turned and hobbled back into her hut made of sticks.

"Oh, why doesn't anyone answer direct questions around here?" Marissa scoffed.

Merlin looked at her and meowed.

"Silas!" called the old magician as he stood outside the door of the small wooden cabin. He rapped on the door with the head of his walking stick and called again. "Silas!"

The door opened and a middle-aged man appeared. He was similar in stature to Marcellus; the same height, body type, long gray hair, but was not yet stooped with age. He also didn't wear the long flowing robes of the old magician, but wore instead loose-fitting leggings made of rough brown cotton, and a gray tunic that had a rope tied around the waist as a belt. Upon his feet were black leather boots, knee high, scuffed and faded from use. When he spied Marcellus, his eyes narrowed in suspicion.

"What brings you to my door, old man?" he asked darkly.

"Oh, you could not foresee this visit?" Marcellus said with just enough sarcasm to make Silas' cheeks redden.

"You know I cannot see into your activities," he snarled. "How would I know of this…visit?"

"Yes," Marcellus agreed. "Wizards cannot penetrate other wizards' auras. I suppose that is why I could not foresee your attempt at murdering the queen's ward."

Silas raised his chin slightly in a defiant attitude. "I have no knowledge from whence you speak," he said smugly. "To what possible end could I have reason to murder anyone?"

"To the possible end of becoming Court Wizard," Marcellus answered.

Silas waved his hand in a dismissive gesture, and made a scoffing sound. "I have no desire to become Court Wizard," he said. "I have no use for the queen or her court. It is too much of a bother to be a part of that world."

"Then why did you send the fire?" Marcellus asked him matter-of-factly.

"Fire?" Silas repeated. "What fire?"

Marcellus could tell Silas was lying. Even if he didn't send it, Silas knew of the fire. But he decided to play along.

"You do not know of the fire?" he said casually. "The one that was in the child's chambers and almost claimed his life?"

"How would I know of that?" Silas asked evasively.

"Again, your powers are not great enough to see things?" Marcellus said, his voice holding just a hint of a mocking tone. "I would have thought you could have at least seen the fire itself. After all, the queen herself rescued the little tyke."

Silas's eyes widened slightly in surprise. "She put her own life in danger to protect an orphan child?" he asked. "The queen should not risk herself for someone so insignificant."

"Not being at court, you would have no way to know that Her Majesty sees no life as insignificant. She will put herself in danger to protect her subjects every time the need arises."

"That is foolish," Silas said. "The queen is much too valuable to put herself in harm's way. Should something happen to her, who would ascend the throne and rule Avonridge?"

"I have wondered that myself," Marcellus said cryptically. "Perhaps that responsibility would fall to someone with great knowledge of the workings of the kingdom, someone with a special talent to assist him."

Silas raised his eyebrows in disbelief. "And you think you are the best candidate?" he asked. "And just what would make you think you have the necessary qualifications to become king?"

"I do know how the court is run," he answered lightly. "I could easily keep things moving along at a steady pace. Should something happen to the queen, I could step into her position without so much as a ripple in the surface of Avonridge."

"And would you not consider that kind of talk treason?" Silas asked menacingly.

"Treason implies intent," Marcellus said. "I am only commenting on the possible outcome of events beyond my control."

"Beyond your control?" Silas repeated. "Perhaps not so beyond your control. You yourself said the queen would risk her life to protect her subjects. Perhaps you knew that if you sent the fire after her ward, she would rush into the flames to save him."

"Perhaps before she proved her worth, that may have been true," Marcellus said. "But I have since acquired a respect for the queen. She is quite capable. I have no desire to supplant her at this time."

"Even so, it would not be advantageous for the queen to perish in an attempt at saving her ward," Silas said thoughtfully. "That would be a detriment to everything."

With that last statement, Marcellus now fully believed Silas had sent the fire after Kevin. He had suspected Silas was only trying to get Kevin out of the way so he could insinuate himself into the court, and now with Silas' concern for Marissa's safety, he was sure of it. He was silent for a few moments.

Finally, Marcellus said, "In addition to the queen protecting her ward, I will be keeping a vigilant eye on him. I have a great personal stake in him remaining safe."

"How could it possibly benefit you whether the child lives or dies?"

"Not only is he the queen's ward," Marcellus said, "but also my pupil. You see, unlike you, I have no jealous feelings for those with powers. I am not intimidated by one with abilities."

"I am not intimidated!" Silas nearly shouted, and then quickly composed himself. "I have the utmost respect for those that are like you and I. I would have no reason to try to cause harm to anyone with abilities." But his countenance remained anxious.

"No, I am sure you would not. Just let me leave you with a small bit of information. I have grown quite fond of the child. I foresee a long and enjoyable relationship with him as he grows in the knowledge and control of his abilities. I would take it as a personal attack should anything happen to him. A personal attack that would require retaliation."

Silas paled slightly, but insisted, "As I said, I would have no reason to try to harm the child. I have no desire to be involved in life at court."

"Good!" Marcellus said, and nodded. "Then we are through here." And he took one step backward, and vanished before Silas' eyes.

Silas gave a small gasp of surprise, and then muttered to himself, "I simply must learn that trick."

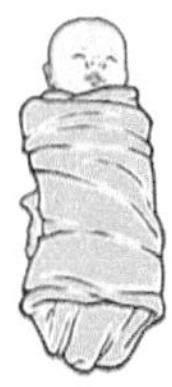

FORTY-SIX

Marissa was true to her word and kept Kevin tight by her side. She had a small bed brought into her chambers so he would even be close when he slept. The ladies-in-waiting took it with mixed reactions that they wouldn't be needed to look after him, at least for the time being. Some had grown genuinely fond of the boy. But Marissa was adamant, and didn't let him out of her sight.

The members of the court seemed not to notice that the child was always at Marissa's side, as most knew how much she loved him. They had seen him in her company often enough to find it not at all out of the ordinary that he was with her continually.

Sitting on her throne in the great meeting hall, Kevin playing with his carved livestock beside her on the dais, Marissa spied Jasper hurrying toward her, his head continually bobbing up and down in reverence to his queen. Always and forever vigilant, Sir Erick stood off to the side, and Ruth was also in attendance.

Jasper stopped about twenty feet in front of Marissa, and bowed at the waist. "Majesty," he said, and waited for her to answer.

"Yes, Jasper," she said easily. "Come forward and tell me what's on your mind."

He took several steps toward her, and stopped at a discrete distance.

"Majesty," he said again. "There is a man at the doors of the palace who wishes to see you. He carries a baby."

"Oh, dear," Marissa fairly moaned. "Not another orphan," and she rose from the throne, took Kevin's hand, and started toward the front of the palace, Jasper obediently falling in behind her.

Ruth hurried to catch up, and called to her, "Majesty, please wait."

Marissa stopped and turned toward her friend. "Yes Ruth, what is it?"

"Ah, yes, Majesty," Ruth stammered. Hesitantly, she said, "May I have a quick word with you?"

"Of course. What is it?"

"Well, it is just, I mean…" She was clearly anxious about what she wanted to say.

"Ruth," Marissa said gently. "You are my friend. Please feel free to speak your mind."

"Yes, Majesty," she said, somewhat relieved. "Well, Majesty, when Kevin was brought to you, you very graciously took him as your own. Do you think it would be prudent for you to do that with another child?"

Marissa laughed gently. "You're absolutely right, Ruth. I have no intention of adopting another child. But I made a promise to my subjects that I would take care of each and every one. I will not let the baby be neglected."

"I understand, Majesty. But you have such a soft heart. I am just worried you might take on more than you can handle."

Marissa smiled at the young woman. "Don't worry. I am not going to do anything foolish. Why don't we wait and see what the situation is before we make any judgments?"

"Of course, Majesty," Ruth said, and they continued to the front doors of the palace.

Standing on the stone landing was a man in a rough tunic and burlap leggings, carrying an infant. He looked anxious, exhausted, and a whole host of other things Marissa couldn't immediately identify. When he spied the queen, the man rushed forward, and dropped to one knee before her.

"Majesty," he said in a hoarse voice. He continued to kneel, not saying anything further.

Marissa waited a moment, but it soon became evident he was not continuing.

Ruth said, "Come now. Her Majesty does not have time for you to find your tongue."

"Yes, yes," the man stammered, still kneeling. "Majesty, we have been told you will not let a baby go abandoned."

"That is true," Marissa said. "But please get up off your knee. It is all right for you to stand and talk to me."

Obediently, the man rose, keeping his head down and not making eye contact.

"Now, is this baby an orphan," she asked, leaning forward to pull the rough blanket back to look at the child, and was surprised to see it was a newborn.

"Well, not exactly, Majesty," he said. "Her mother, my wife, died birthing her."

Marissa could hear great pain in the man's voice, and her own heart ached for the baby, the mother, and this man who had lost someone dear to him.

"I am so sorry," she said. "It must be horrible for you."

The man nodded in agreement, and Marissa saw a tear fall onto the baby's blanket.

"So, Majesty, since she has no mother," the man's voice grew hoarser, "she is in need of a home."

The baby began to fuss.

"But, are you sure you want to give up this child, the child your wife gave you?" Marissa asked. "You will be losing a second loved one. Are you sure?"

The man was silent for a moment, then again nodded. "I must, Majesty. My sons and I cannot care for her. They are almost grown, ten and twelve years old, and must tend the livestock and crops. No one can be there for her. She needs you, Majesty. Please."

Now the baby started softly crying.

"What is your name?" Marissa asked.

"Calvin, Majesty."

"Calvin, I will make sure she gets a loving home," Marissa assured him. "I will make sure she is well looked after, and given everything she needs."

Calvin nodded again, and stepped forward to place the crying baby in Marissa's arms.

"We are grateful, Majesty," he said. "I knew I did right by coming to you."

Marissa took the baby and cradled it against her breast. "What is her name?" she asked.

"She has no name, Majesty."

"No name?" Marissa said. "Why have you not named her?"

"Because I cannot keep her," he replied stiffly.

Marissa nodded sadly, and asked, "When was she born, Calvin?"

"Last eve, Majesty. And she has not had much to eat. We had very little milk."

"Ruth, do we have someone who can act as a wet-nurse?" Marissa asked.

"Yes, Majesty," she said, and turned to Jasper. "Boy, go fetch Shandra."

"Yes ma'am," he said, and immediately raced down the palace steps on his quest.

The baby's cries grew in volume, and Marissa, not being familiar with infants, was amazed at the strength she had in her lungs.

Calvin wiped the back of a hand across his eyes, bowed, and said, "Thank you, Majesty." He then turned and hurried down the palace steps toward the gate.

FORTY-SEVEN

Marissa spent as little time as possible with the infant, not wanting to fall in love with her. She was determined that she would not take this one as her own as well, but was concerned that she got all the love she needed. Shandra nursed her, and an older lady-in-waiting, Danielle, took her immediately to her heart. Danielle had spent her whole life being a lady-in-waiting, and had never regretted it, but once in a while she wondered about the road not taken, that of a wife and mother. So, for at least a little while, she would try the road of motherhood.

Marissa decided the day the baby was brought to her that she needed a name. She thought about Alice, the first queen of Avonridge, but rejected that, not wanting to put undue pressure on the baby. Marissa remembered back to her life before Avonridge, which was an almost impossible task anymore, and thought about the old 'I Love Lucy' reruns she used to love. She had always liked the name Lucy, and so decided to bestow it upon the infant. From the day she came to the palace, the mother-less infant was Lucy.

Lucy was doing well. Between Shandra and Danielle, she never lacked food or attention. When she cried, either one of the women was there to comfort her. But Marissa noticed after about three days that the shine seemed to have worn off for Danielle. The constant attention Lucy needed was completely alien to Danielle, and it quickly grew tiresome. Marissa knew something needed to be done soon.

As luck would have it, awful, heart-breaking luck, there was an earl and his wife due to give birth at any moment. The court midwife was called to their manor house in the middle of the night on the fourth day after Lucy came to the palace. The midwife was skilled and capable, but nature has a will all her own.

Even though the birthing went smoothly, and it was fairly easy for the mother, the baby was born with the umbilical cord wrapped around her neck, and did not survive the delivery. The earl and his wife were devastated.

By morning, the news of the tragedy had reached Marissa. Immediately she knew what to do, and donned her traveling cloak to shield against the morning chill. She called for Shandra to bring Lucy, and with Kevin by her side, the four made their way to the earl's manor house.

Marissa was not sure how the grieving couple would react to her proposal, but she knew she had to try. After all, it was such a coincidence, it seemed meant to be. When the little group arrived at the manor house, the male servant admitted them. Standing in the main room of the front entrance, Marissa's heart ached at the sobs she could hear coming from up the stairs.

Such a feeling of grief encompassed the room it was almost suffocating. The servant that had admitted them, Samuel, was walking stoop-shouldered, and had tears in his eyes.

"Samuel," Marissa said gently. "This is such a tragedy. I am so truly sorry."

"Thank you, Majesty," he said, and a sob caught in his throat.

"Tell me, what's happened to the infant?" she said.

"The master was overcome with grief, Majesty, and he knew the mistress would not be able to bear the pain of seeing the dead child." He paused to collect himself, and then he continued. "Just an hour after the birth, he took the babe out to the family cemetery and buried it with a small spade and his bare hands, all the while crying with a broken heart."

"How dreadful," Marissa whispered.

"Truth to tell, Majesty, when I saw you with the woman carrying this child, for a moment I wondered if God had changed his mind and brought the master's babe back."

"Yes, that would be a most welcome miracle, Samuel." Then she touched his arm and said, "I hate to disturb them at this sad time, but I must see them."

Not wanting to have the earl leave his wife, Marissa bade Samuel bring her to them. He obediently led the queen and her company up the stairs to the sleeping quarters.

At the door to the room that housed the crying woman, Marissa had Shandra and the baby wait outside with Kevin. Not wanting to burden the grief-stricken couple any more than necessary, she entered the room alone.

Curtains were drawn over the windows, and a single candle burned on a sideboard on the opposite wall from the door. The room was dark and still, the only evidence of occupancy being the brokenhearted sobs of Belinda, the young mother without a child.

A large, oak, four-poster bed stood against the wall to the right, and a young woman lay crying, propped up on pillows, covered with a soft beige blanket. Her dark hair was matted with sweat and tangled from hours of labor, and she wore a white satin dressing gown. A man in his late twenties sat on a chair next to her, holding her hand in both of his, hunched at the shoulders in obvious pain.

When Marissa stepped through the door, the earl, Richard, looked up from where he was sitting beside the bed.

"Did I not tell you not to disturb us," he growled through tears. But when he saw who it was that disturbed them, he immediately stood and said, "Majesty, forgive me."

"Richard, forgive me," she quickly replied. "I know I am intruding on something so very personal. My heart is breaking for your loss."

He didn't speak, but nodded in response. Belinda tried to choke back her sobs, but failed, and continued to cry.

"Richard," Marissa said, "sit and comfort your wife."

He sat, and took Belinda's hand again.

Marissa tentatively stepped closer, not wanting to infringe upon their grief, yet wanting to help ease their pain. She wasn't sure if her idea would work, but had to give it a try. Softly, she stepped closer still, and gently laid a hand on Richard's shoulder.

"I cannot say I know your pain," she said. "But I know it is immense."

Belinda looked up at Marissa, and Marissa's heart ached anew at the torment in the young woman's eyes.

"I also know," Marissa continued, "that you feel like it will never subside."

Belinda lowered her eyes and nodded in agreement.

"I've been told there is no greater pain than the loss of a child," Marissa said. "And I can see in your eyes that this is true. I want to take the pain away and leave you with peace in your hearts."

"If only that could be possible," Richard choked miserably.

"If I said I may be able to ease some of your pain, certainly not all, but perhaps a small amount of it," Marissa offered, "would you be willing to hear me?"

"As my queen," Richard said, "I would listen to your words. But there is nothing to be said that could calm this burning pain."

"All I ask is that you listen," Marissa said.

Richard nodded, and Belinda, having finally quieted her heartrending sobs, looked up at Marissa.

"I have never been one to believe in coincidence," Marissa began. "I have always believed things happen for a reason. We do not always know the reason. In fact, more times than not it remains a mystery to us. But we must trust in God's wisdom."

"God has abandoned us, Majesty," Richard said, and Belinda nodded silently.

"I know it feels so now," Marissa agreed. "But trust me when I tell you it is not so. For some reason we will probably never know, God has decided to take away from you something very precious. But he has not abandoned you to grieve forever. He has brought you something equally as precious." She paused, waiting to see their reactions.

Through her pain, Belinda became curious. She looked at Marissa questioningly, at least willing to hear her out. Richard turned to face Marissa, somewhat curious as well, and spoke.

"What do you mean, Majesty? What could God possibly bring us that is precious?"

Instead of a direct answer, Marissa said, "A couple of days ago a man came to my palace. He too had a heartbreaking tale to tell. His sons are older, and need to work the farm. His wife began to deliver their next child, a girl, but something went terribly wrong. As the mother brought forth the child, a tragedy struck, and the mother did not survive. The baby girl had no one to care for her. That is, until now."

Marissa waited for the young couple to fully comprehend her words. When understanding became evident, Belinda sucked in her breath.

Richard shook his head in disbelief, and said, "Do you mean…" and he paused, unable to continue.

"Yes, Richard," Marissa said. "There is an infant in need of a mother, and here is a mother in need of an infant."

Belinda's eyes again filled with tears, and she spoke for the first time since Marissa had entered the room.

"Majesty, are you truly saying…" and she too seemed at a loss for words.

"I know your own child can never be replaced," Marissa said. "But would you be able to find it in your hearts to take another baby as your own?"

Richard looked at Belinda, who looked at Richard.

"The decision is completely Belinda's" he said, and waited for her to respond, his eyes never leaving the distraught face of his wife.

"I don't know," she said softly, confusion, expectancy and many other emotions coloring her cheeks.

"Why don't you meet the baby," Marissa said gently, knowing, since they were not averse to the idea, that would do the trick. "Shandra," she called, and the young woman stepped through the bedroom door, carrying the infant, Lucy, little Kevin close by her side.

Marissa beckoned Shandra closer. The nursemaid walked over to where Marissa stood next to the bed. She placed the baby in Marissa's arms, and turned and walked back to stand just inside the door.

Marissa pulled the blanket away from Lucy's face and folded down the edges so the earl and his wife could get a good look at her. Belinda folded her hands and tucked them under her chin, almost as if afraid to reach for the baby. Her face, however, was expectant as Marissa brought the baby close.

"Belinda," Marissa said gently, "this child has no mother. Just hold her for a few minutes," and she leaned down and offered the baby to Belinda.

Belinda hesitated only a moment, and then opened her arms to receive the baby. Marissa laid Lucy in Belinda's arms and stepped back. Belinda gathered the baby to her bosom, her eyes never leaving the infant's face.

Richard leaned close and tentatively reached to touch the child. Belinda looked up at her husband, and the shadow of a smile could be seen on the young woman's lips.

"Richard," she whispered. "This is God's design."

The earl brushed his fingertips over the baby's cheek. Lucy lay quiet in Belinda's arms, looking content. Richard looked up at Marissa.

"Majesty, you have given us the greatest gift we could ever receive," he said. "You have given back to my wife a reason to continue living. I had feared for her life, Majesty. And now I need not worry. How can we ever thank you?"

"By never letting this baby feel like she does not completely belong to you," Marissa said.

"This child will never have cause to believe she is not loved by a true mother and father," he assured her.

"Then I have done what I came here to do," she said, and left the little family to heal.

FORTY-EIGHT

The magician Silas was again summoned to his front door by the knocking of a visitor. Forsythe stood at the door of the cabin, an angry and anxious look on his face.

"I have need of your services," he said abruptly, and the magician's eyes widened.

"Again?" Silas said, somewhat contemptuously. "Did you not sufficiently frighten the queen on your first attempt?"

"No, I did not sufficiently frighten the queen," Forsythe growled. "She fought off the wolves on her own. She needed no assistance, and indeed made them flee in fear of her. I only succeeded, therefore, in bolstering her confidence."

"My, my," Silas said. "What a pity it did not go as planned. Perhaps you should abandon your plans and settle for marrying another."

"So, you know!" Forsythe said angrily. "You know of her plan to have me wed to one of her ladies-in-waiting."

"I know of most things at the court of Avonridge." Silas replied easily. "After all, I have plans of my own."

"Yes," Forsythe said. "You plan to become the court wizard. How quaint."

"And why should I not?" Silas challenged. "I would be given chambers in the palace. I would be the queen's advisor."

"Why would you want all that?" Forsythe asked. "You claim to have no interest in life at court. Why have you changed your mind? And

336

Marcellus doesn't abide in the palace. Why would that be something the queen would even want?"

"Marcellus is a fool, entrenched in the old ways. For me, the winters grow colder each year," Silas grudgingly explained. "It is true. I never had any interest in the goings on at the palace. But as I grow older, my bones protest more and more at the cold that infiltrates my cabin."

"But to what purpose do you send the fire after the child in Marissa's care?"

"If the queen," and he said the title forcefully, "knew she had a court wizard just waiting to grow to adulthood, she would have no need of me. I must eliminate the competition before she grows too overly fond of him."

"You are too late for that," Forsythe said. "She grew too overly fond of him the day he arrived at the palace."

"Then more's the pity," Silas said. "She will grieve when the child is finally removed from my aggravation. Too bad, I had wanted to save her the pain. After all, I bear her no ill will."

"Nor do I," Forsythe said, a little too quickly. "I simply wish for her to see that she needs a man to care for her."

"Is that so?" Silas asked. "Or has that changed? After all, she is ordering you to marry another. Could it be you want the wolves to eliminate the queen before she forces you into a marriage you clearly do not want?"

"Of course not," Forsythe said, but his eyes took on a hooded appearance. "I could never intend harm for Marissa. I simply want her to see that she needs me before I marry Bethany."

"Good," Silas said, but suspicion colored his countenance. "That is good, because there is another who is ready to step in and fill the void should the need arise from the death of the queen."

Alarm sprang to Forsythe's face. "Oh," he said, trying to feign calm. "And who would be treasonous enough to say they would ascend the throne upon Marissa's death?"

"That is unimportant," Silas said. "What does interest me, however, is your familiarity with your queen. You simply refuse to refer to her by title, and call her by her given name. That implies a closeness that, forgive me but, I just do not believe is there."

Forsythe's cheeks reddened. "Do not concern yourself with the closeness between Marissa and myself. That has no bearing on our arrangement. Besides, if you had succeeded in frightening her with your pack of wolves, this closeness would be evident to the entire kingdom by now."

"Ah yes, then she would have run straight into your waiting arms for comfort and protection."

Ignoring the sarcasm, Forsythe said, "Exactly. However, you failed."

"My wolves did as they were commanded to do," Silas countered. "They awaited the queen outside the Vulture's Claw and attacked. They were given strict instruction not to harm the queen, and they obeyed. How could I know she would be such a formidable opponent?"

"Perhaps if you had some knowledge of life at court," Forsythe scoffed, "then you would know she bested King Sorens."

"I am fully aware of that," Silas assured him.

"Then you should know how formidable she is, and you should have planned accordingly."

"Sometimes things do not go exactly as planned." Silas explained patiently. "One must use trial and error to eventually succeed."

"Trial and error," Forsythe repeated derisively. "Like your error in eliminating the child while he still lived in his parent's cabin?"

"Yes, quite a pity that," Silas said sadly. "I would think it worthwhile if the fire had succeeded in its primary quest. But to overcome the entire family save the intended victim, too bad indeed. What a waste."

"And then your error in eliminating him in his palace chambers," Forsythe goaded.

Silas' eyes narrowed in contempt. "Again, yes! I did not know the queen would put herself into such danger to protect the child."

"Two errors, or should I say failures?" Forsythe said smugly. "Perhaps your method of attack needs to be rethought."

Silas nodded, anger turning quickly to thoughtfulness. "Indeed, perhaps we can both attain our desires with one more ploy."

"You have my attention," Forsythe said.

"Perhaps," Silas began, "you could achieve your goal of the queen needing you for protection, and I could achieve my goal of eliminating my future competition with one decisive attack."

"Go on," said Forsythe.

"My wolves do as they are instructed. The next time the queen and her ward are out for a leisurely stroll…"

"If Marissa witnesses the bloody demise of her little bundle of joy, she would be devastated." Forsythe said, with just a hint of sarcasm.

"She would need comforting," Silas said.

"And she would learn that she cannot handle all the burdens of palace life on her own," Forsythe added. "Yes, I like the idea."

"As with your first use of my wolves, you must go to the location at which you want the…" he hesitated, choosing his words, "incident to occur."

"I remember," Forsythe said. "That is a most inconvenient requirement. I had to get close to the Vulture's Claw for the first… incident. So close in fact, that my waistcoat was torn. When I returned to my home, I noticed a piece of cloth missing. I do not like the idea of putting myself that near danger."

"Be that as it may," Silas said, "it is necessary."

"How do I know when Marissa will be out walking with the child? And where?"

"That you must figure out on your own," Silas said. "If you desire this enough, you will make it happen."

* * *

"Majesty, I simply do not understand it," Roland said. He had taken to pacing back and forth before Marissa in frustration. "King Sorens absolutely refuses to grant anyone from your kingdom an audience. He knows we hold two thousand of his men, yet he seems completely against negotiations."

"I don't understand either, Roland," Marissa said. "What could possibly be his reasoning for abandoning his soldiers?"

"I cannot venture a guess, Majesty. At first, he would not meet with the envoy, claiming he did not grant audience to lowly messengers. The second envoy was my second-in-command, but again King Sorens refused him. Now I, myself, your Captain-of-the-Guard, have requested audience with him to negotiate the return of his soldiers, and again he refuses. Could he possibly be expecting the queen herself to engage in simple negotiations?"

"He can expect it all he likes," Marissa said casually. "But if he sincerely thinks I would willingly walk into his kingdom, after he attacked mine, then he has lost his faculties."

Roland nodded in agreement. "I would never allow that, Majesty. He is no doubt still smarting from the thrashing you laid upon him. He would most definitely do something contemptible to try to harm you, or even abduct you and hold you in exchange for your kingdom. No, you must not put yourself in that sort of danger."

"Then I am through making attempts, Roland," Marissa said. "If Sorens wants his soldiers, then he will have to make the effort. I will wait for him to come to me."

"Very good, Majesty. But in the meantime, what do we do with the two thousand men?"

"Start bringing them into life at Avonridge. I will address them. They have already been of help in keeping the grounds in order, haven't they?"

"Yes, Majesty," Roland said. "They have been put to work on more than one occasion cleaning the snow from the streets. They have all seemed quite content to be working, instead of simply sitting the time away in the barns. After all, while they were idle, problems would break out among them, and there were a few fights."

"Yes, being cooped up like that with no visible end in sight is quite frustrating. Having them perform menial labor is at least a release for that frustration. Have them gathered in the main barn for me, Roland. I'll speak to them now."

FORTY-NINE

The group of captured soldiers sat in the large barn, talking among themselves. Marissa had donned her gold tiara and royal purple cape, giving the impression of obvious royalty, but not being garish by wearing her heavy crown. For this occasion, she was without Kevin. The only time she was comfortable not having him in her presence was when he was with Marcellus, who was just as protective of the little boy as Marissa herself, and now the two were engaged in Kevin's lessons.

When Marissa entered the barn, the soldiers became silent. She walked to the front of the large group, marveling at the sight of two thousand men sitting before her. Not one rose to his feet in respect, and Roland was about to shout the order to do so, but Marissa gave a quick shake of her head. He relented, but was obviously anxious.

Positioning herself in front of the men, and raising her voice to be heard by all, Marissa said, "Gentlemen, I wish to speak to you. As I am sure you know, I am Queen Marissa of Avonridge. The manner in which you all came to be guests in my kingdom was not a pleasant one," and she heard several indecipherable grumbles throughout the crowd.

"However, I have tried to make your stay here as tolerable as possible. I have supplied extra blankets on the coldest nights, and provided fires to warm the barns. I hope you have found the food I offer acceptable, and I have even offered some recreation in the form of dice and other games."

Here and there she heard a word of agreement from the soldiers.

"Now, I understand," she continued, "that this is not the ideal position for men of action such as yourselves, and therefore I am willing to negotiate some changes to your present situation."

"Here, here," was heard throughout the crowd.

"Your King Sorens has shown himself to be my enemy by attacking my kingdom," she said. "There are two directions to take in regard to you, his soldiers. First, I could declare that, since you are soldiers of Sorens and Southcross, you are also my enemies, and deal with you in the manner befitting my enemies. But I prefer the second direction. You were simply loyal men doing your king's bidding, and hold no personal ill will toward me or Avonridge."

She paused for a moment, and saw several men nod in agreement.

"Now, as far as that loyalty is concerned," she said, "does it not deserve loyalty in return? Do you men, loyal subjects of Southcross, not deserve your king's protection?"

"Yes, Majesty," one man said from the front of the crowd.

"Yes," she agreed. "If my soldiers were captured and held by another monarch, I would do everything in my power to ensure their safe return. However, your King Sorens does not seem to hold those same feelings. Not only has he not sent an envoy to negotiate your release, but he has refused to grant audience to all three of my envoys."

"That is a lie!" shouted someone in the crowd.

"Hold your tongue or you will be punished," roared Roland in fury.

"But that cannot be true," ventured another, almost pleadingly.

"It is true," Marissa said gently. "Think about it. I have treated you well, fed you sufficiently, kept you as warm as possible. What possible reason could I have to not return you to your home?"

There was silence for several moments, and then one man asked, "Is it so, Majesty? Does King Sorens leave us without a second thought?"

"It seems so, soldier," she said. "I have not been able to understand his refusal of my envoys. He simply will not negotiate."

"Then to hell with Sorens!" shouted a man in the crowd.

"Hold your tongue, soldier," yelled another. "This may still be a ploy."

"Think about it," said a third. "What could Queen Marissa possibly gain by lying to us?"

The men began to debate back and forth the legitimacy of Marissa's claims, and as the discussion went on, it began more and more to lean in Marissa's favor. There were several challenges for a convincing argument that Marissa was lying, but none could be heard coherently.

The fact that the soldiers had been captive for several months also lent credence to Marissa's claims. She let them talk back and forth, and soon they all seemed to accept that their king had abandoned them.

One man stood up in the crowd. He seemed to hold some rank, because the crowd quieted and gave him their attention.

He addressed the crowd, and said, "It seems we have been forgotten by our king. He does not try to negotiate for our return, and therefore has left us for dead. I say we hear what Queen Marissa has to say."

There were murmurs of consent throughout the crowd. Not one protest could be heard.

"Thank you, soldier," she said. "I have an offer for you, all of you, that I will give you time to consider. Talk it over among yourselves and let me know what you think."

She noticed some men nodding, and got a strong feeling that at least they would listen to her offer.

"Men of Southcross," she said loudly. "I hold no animosity toward you for your king's aggression against my crown. You have behaved in an acceptable manner since you have arrived in Avonridge, and therefore I extend to you an invitation to consider becoming subjects of Avonridge."

The men were silent for several long seconds, until finally one man in the crowd stood up and said, "I have no family to return to in Southcross. It makes no difference to me who I call sovereign."

"Where is your loyalty to your king?" another man called, but the accusation in his voice was almost nonexistent.

Another man said, "I have a wife and son in Southcross. I do not want to leave them."

The first man said, "But if King Sorens does not try to bring us home, what good is it to waste your life wishing for something that will never happen?"

Marissa held up a hand for silence, and the soldiers obeyed. "Men, you may take as long as you wish to discuss my offer. Please keep in mind you do not all have to agree. I will open my kingdom to those who wish to stay, and make other arrangements for those who do not. Do not be afraid of any retaliation should you decide not to become loyal to Avonridge. There will be no repercussions if you refuse my offer."

"And a generous offer it is," Roland growled loudly. "Her Majesty could simply allow you all to waste away waiting for your king to act on your behalf. But she has given you the chance to start a new life in a prosperous kingdom. You would do well to accept her offer."

"Thank you, Roland," Marissa said. "But they are free to make their own decisions." Then to the crowd, she said, "Send word when you have your answer. I will be waiting." And she left the makeshift prison.

Marissa put a light jacket on Kevin and donned a hooded cape to go walking in the cool air of late spring. The day was bright and the air clean and crisp as they stepped out onto the palace steps.

Sir Erick wanted to accompany the two, but Marissa insisted they be able to walk alone. They were going nowhere in particular, and simply wanted a leisurely stroll through the lower meadow to see the new spring vegetation. This would be Marissa's first spring in Avonridge, and she wanted to see what the beautiful land had to offer. In fact, Forsythe had suggested she would enjoy the lower meadow, as it was so full of new growth. She assured Sir Erick they would only be an hour or two, and off they went.

The spring grass in the lower meadow was sweet and green and knee high to Marissa, making it almost up to Kevin's chin. It was difficult for him to walk through it, so Marissa picked him up and put him on her shoulders. She wasn't sure if he'd allow that, since he was so independent, but with the difficulty he had walking through the grass, and the bird's eye view he now was witnessing, he was quite content.

As she walked along with Kevin on her shoulders, Marissa pointed out the different flowers. She didn't know the names to most of them, but she showed Kevin the pretty purple flowers, and the pretty red flowers. At one point, she stopped and picked a sprig of Queen Anne's Lace and tucked it into her hair.

The cool spring day was wondrous, and Marissa and Kevin were fully enjoying the fresh smell of clean air. Suddenly, however, Marissa felt Kevin stiffen.

"Mama," he said tentatively.

"Yes, sweetheart. What is it?"

"Mama," he said again, and this time Marissa could hear fear in his voice.

She craned her neck to look up at him, and saw the direction he was looking. He was staring a little off to the left, and she looked in that direction. She could see nothing out of the ordinary, but decided to trust the little one's instincts. She took Kevin off her shoulders, and turned to head back to the palace, carrying him on her hip.

Kevin was becoming more agitated, and said, "Mama, run."

Becoming alarmed herself, Marissa began to trot through the tall grass, all the while checking the meadow in the direction that Kevin was pointing. After several minutes, she thought she could see things popping up and down in the grass, way out in the distance. Immediately she knew the things were wolves bounding through the grass, leaping over the tall blades as they ran.

"Mama, run," Kevin said again, and Marissa moved as fast as she could.

Running through the grass, she cursed the long skirt of the gown she wore, wishing instead for the denim jeans she used to wear before she came to Avonridge. But unfortunately, those were a lifetime away, so she did the best she could with what she had.

It only took her a few moments of running to realize the wolves would cut her off long before she reached the safety of Avonridge proper. Panic began to fill her heart as she thought about little Kevin, and realized she had absolutely no weapon with which to defend him. Now she cursed herself for being so arrogant. Avonridge was a peaceful kingdom filled with magic and wonder, but it was still a wild place with ordinary dangers.

The wolves closed the distance between them quickly, and they fanned out around Marissa and Kevin in an arc, causing the queen to stop running. She stood still as the five wolves from her past encounter encircled her and her precious little boy. She spotted the lead wolf, and smiled in spite of herself when she saw the scarred-up slashes that Merlin had inflicted. Desperately she hoped the little cat would rescue them again.

The lead wolf growled, and the others did the same.

"Mama," Kevin whispered, and she instinctively held him tighter.

"Mama," he said again, more insistently. Marissa looked at him, and he returned her gaze, looking directly into her eyes.

"Mama, I want down," he said calmly.

Marissa shook her head 'no', and said, "Absolutely not, little one."

But Kevin insisted. "I want down," he said, and pushed against her shoulder.

The lead wolf growled again, and with his head slung low between his shoulders, he took a step toward them.

"Mama, I want down!" Kevin nearly shouted. Deciding to trust this little spitfire, she gently put him on the ground.

As soon as Kevin's feet landed, the lead wolf leapt at them. Marissa tried to gather Kevin back into her arms, willing to protect him with her own body, but Kevin swung his fist at the wolf as if to throw a boxer's punch. The wolf, still about five feet away, was stopped in midair, and thrown back about thirty feet. He landed with a loud thud and a yelp of pain.

Another wolf leapt at the queen and her little boy, and again Kevin punched the air in its direction, sending it flying backward through the air. Suddenly Marissa was hit from behind and driven down to the ground with a third wolf standing on her back.

Kevin turned toward the wolf that attacked his mama, and in fury screeched, "Get off!" and threw both fists at the wolf, who was yanked straight up by invisible hands. The wolf hung suspended for several seconds, then flew off to the side as if thrown by an angry giant.

The wolves didn't seem so easy to frighten this time, however, and all five regrouped and resumed their circle around Marissa and Kevin. Marissa got up off the ground and reached for the little boy, needing to keep him close. She wondered why they hadn't run yet, and then remembered that Merlin was the driving force behind their fleeing at her previous encounter.

Desperately she wished again that he would come to her rescue, and thanked God himself when she saw the little black and white ball of fluff sail through the air to land on the lead wolf's head, screeching and hissing. This time, however, the wolves seemed ready for the assault, and one ran to the lead wolf, grasped Merlin in its teeth, and yanked him off the wolf's head, throwing the little cat to the ground at Marissa's feet.

Immediately Merlin jumped up and prepared to attack again, but Kevin said, "No Merlin. I fix this." Kevin grabbed Merlin around the middle, picked him up and hugged him to his chest. Kevin then grabbed Marissa's hand, and shouted, "Vanish!" just as the lead wolf leapt again.

Marissa didn't notice anything at first, just Kevin tugging her out of the path of the leaping wolf. They simply sidestepped the wolf's assault, and stood still. The wolf landed and looked around, as if completely confused.

The other four wolves looked around as well. Marissa looked down at Kevin and Merlin, and was shocked when she couldn't see them. She could feel Kevin's hand in hers, but could not see either her precious boy or her beloved cat. Slowly she realized she also couldn't see the skirt of her gown. She looked down at her body, and could not see that either.

The predators began milling about as if searching for elusive prey. They walked back and forth, sniffing the air and the ground. Marissa stood statue still, not completely at ease with this method of defense, but thankful for it just the same.

Slowly the wolves fanned out, continuing to sniff the air and ground, obviously finding no trace of their intended victims. Marissa felt a tug on her hand, and allowed Kevin to lead her out of the meadow.

FIFTY

Marissa waited in her sitting room for Roland. She had a mission for her Captain-of-the-Guard, and was anxious for it to be done.

Kevin sat at the table, as usual not out of her sight, playing with his carved animals. Again, he seemed no worse for wear after the attack. He looked up at her and smiled.

Nodding at her little bundle of joy, she said, "So, you've come a long way in your studies. You are quite the little magician, aren't you?"

Kevin giggled a precious giggle, and said, "Marcellus teach me."

"Yes, he has taught you well," she agreed. "When you started playing tricks on the ladies and me, I got annoyed. But now I am grateful you know those tricks." She reached down and hugged him, and said, "You saved our lives. All three of us. And ultimately, I have Marcellus to thank for that."

Kevin went back to happily playing with his animals, and in a few moments, Roland walked in the door.

"Bowing at the waist, he said, "You sent for me, Majesty?"

"Yes," Marissa said. "I have a mission for you."

"I am at your service, as always, Majesty," he assured her.

"We have a menace running around the kingdom," she said. "There is a pack of wolves out there, causing havoc. I myself have run into it twice now."

"A pack of wolves?" Roland echoed.

"Yes, and they are quite dangerous."

"I know wolves roam the forests of Avonridge, but I have never heard of any attacking anyone."

"Well, you have now," Marissa said. "I have been attacked twice."

"Attacked?" Roland said, alarmed. "Are you well, Majesty?"

"I am quite all right, Roland," she assured him. "Both times I was assisted in my escape. However, they seem to be growing bolder. I do not want to just assume I will be able to escape them should they attack again. I need you to find them and get rid of them."

"Of course, Majesty," he said quickly. "Shall I eradicate all the wolves of the kingdom?"

"No, that's not a good idea," she said. "I was attacked by five wolves. I believe the same five both times. You should recognize them by the lead wolf. Merlin rescued me from the first attack, and left several large scars on the wolf's head. They are quite visible."

"Yes, Majesty," Roland said. "I will look for that wolf."

"I only want those five removed," she said. "I understand wolves are a necessary part of the natural order. As long as they aren't a real danger, they deserve to be left in peace. But these five do not seem to want peace."

"I understand," Roland said. "I will take a squad of soldiers, and Bruno, the huntsman. He will know how to track and trap them."

"Very good," Marissa said. "I am becoming afraid to walk around my kingdom without a body guard."

"Pardon me, Majesty, but you should not walk around without a body guard. You are the queen, and as such should never be without protection."

"This is a peaceful kingdom filled with loyal subjects," Marissa said.

"Yes, Majesty, that is quite true," Roland agreed. "However, one can never be certain of anything. There is always danger to the crown. You never know when someone might find fault with your court and blame you. Please, Majesty, do not go walking about without protection."

She was silent for a few moments, and then nodded. "I suppose you are right," she said. "There is always the chance that someone may want to harm me."

She didn't tell Roland that she had the overwhelming feeling that indeed someone did want to harm her, and worse, harm Kevin. The fire in Kevin's chambers was proof that the little boy wasn't safe from danger. She couldn't figure out, however, if she too was in danger. The first attack by the wolf pack was on her alone when she had exited the Vulture's Claw. Was the second attack in the meadow aimed at her, or Kevin, or both?

"Let me know when the wolf pack is gone, Roland," Marissa said.

"Yes, Majesty," he said, then bowed at the waist and exited.

"So little one," she said to Kevin, "hopefully soon all this nonsense will be finished."

He just grinned up at her.

"Majesty, I have reason to believe that Silas does indeed wish to harm little Kevin." Marcellus stood as Marissa paced anxiously before him.

"Tell me the details," she said.

"Very little detail," he answered. "But things he said, or rather, the way he said them."

"Did he admit to sending the fire after Kevin?" she asked.

"He admitted nothing," Marcellus replied. "But again, the way he said things, the expression on his face, led me to believe he is in fact responsible. We must keep a vigilant eye on the child."

"That's a given," she muttered.

"Majesty?" he said questioningly.

"Oh, never mind," she said, waving her hand distractedly. "Tell me, Marcellus, what makes you so sure you know his motives? If he hasn't admitted to anything, why do you believe he wants to be the court wizard? He has never tried to be a part of the court before."

"There is no other reasonable conclusion," he answered. "Why would he try to murder a child, not once but twice? Silas' cabin is nowhere in the vicinity of where Kevin's parents' cabin burned. He would have no excuse for interaction with Kevin's family. There could be no other reason for him to try to eliminate the child."

"But why hasn't he tried to get rid of you?" Marissa asked, and stopped pacing.

Marcellus chuckled, amused. "He may wish me out of the way, Majesty," he said. "But he does not possess the power to eliminate me. Silas' abilities may be admirable, but they are not on a level with my own."

Marissa nodded, relieved. She knew she didn't have to worry about Kevin when he was with Marcellus, but this was reassuring, none the less. Then a thought occurred to her.

"Do you think," she asked, "that Silas could have something to do with the wolves?"

Marcellus thought about that for a moment. "I suppose it is possible," he said. "Did you find anything at the scene of the attack that might be evidence to that effect?"

"Nothing that I can remember," she said thoughtfully. Then something occurred to her. "Wait, I did find a piece of cloth snagged in a tree."

"Cloth? Was it rough cotton of the type worn by the forest dwellers?" Marcellus asked.

"No, it was green velvet, like the fine cloth members of the court wear." Suddenly Marissa sucked in her breath. "Forsythe!" she hissed. "It was the same green velvet of the waistcoat he used to wear. The next time I saw him, he had changed his waistcoat to scarlet. You don't suppose he could have summoned the wolves, do you?"

"No, Forsythe has no powers, of that I am certain. However, that brings to mind a possibility just as interesting. Could he be in league with Silas?"

"Forsythe in league with Silas," she repeated. "Could that be true?"

"It is something to keep in mind, Majesty," Marcellus said.

"Well, the problem at hand is the wolves," she said. "I instructed Roland to find the pack and eliminate it."

"That will be a most difficult task if Silas is involved. If they have been sent by a wizard, they will not behave completely in the normal manner. Not only will they be difficult to find, but difficult to eliminate as well. Silas will be controlling their movements, and he will be vigilant of their protection."

"Well, if it can be done, Roland can do it," Marissa said.

Again, Marcellus chuckled, piquing Marissa's interest.

"Is something amusing?" she asked.

"Only this," he replied. "When you were first attacked by the wolves, you were sure I had summoned them. How interesting it is that they seem indeed to have been summoned, but by a different wizard."

"I don't find it funny," she said darkly. "If he did send them, then Kevin and I are not really safe anywhere. Outside he can get to us. In our chambers he can get to us. Where can we go to be completely out of his grasp?"

"Not a place, Majesty," Marcellus said cryptically. "You can only be truly safe as long as you do not let your guard down. You have powers as well. You can use them to combat any threat Silas may send."

Marissa realized Marcellus was correct. She could sense when Marcellus was near, even when no one else could. And she knew something was happening the night of the fire. The only things she hadn't foreseen were the attacks by the wolves, and they couldn't possibly get at her within the palace walls, could they?

"So, since we are pretty much certain Silas has designs on court life," Marissa said, "how do we combat him?"

"As I said before, I will attend to Silas," Marcellus said. "My visit to him was only meant to gather information. Now that I am confident of his motives, I can plan accordingly."

"So, you can get rid of him?" Marissa asked hopefully.

"Perhaps," replied Marcellus. "If that becomes necessary, rest assured I will do everything in my power to rid us of this threat."

"Good," Marissa said. "I do not need the specter of Kevin in danger hanging over my head."

"Nor do I, Majesty," he assured her. "Nor do I."

Marissa had summoned Bethany and Forsythe to her chambers. Bethany had the glow of a woman in love, but Forsythe had taken to wearing a scowl almost continually. To his credit, however, he was civil to his bride-to-be, even allowing her to touch his arm occasionally. Marissa was glad to see he did not pull away from Bethany's touch. Unfortunately, he didn't acknowledge it either.

"So, Forsythe and Bethany," Marissa said. "I have asked you here to discuss your wedding."

At the mention of the wedding, Forsythe's cheeks reddened slightly, but he said nothing.

"This will be the first wedding I will be hosting since my arrival at Avonridge," Marissa continued, "and I want to make it an event to remember."

Bethany bounced on the balls of her feet and clasped her hands to her bosom. "Oh, thank you, Majesty," she said breathlessly. "I so look forward to my marriage," and she cast a loving glance at the dour man that would be her groom.

Marissa was still not completely comfortable with this arrangement. She had thought it such a wonderful idea when she first hatched it, but now was having second thoughts. She had known all along that Forsythe would not be happy, and he didn't disappoint, but now there were other concerns.

Having formed the suspicion that Forsythe was working with Silas, but not knowing their exact motives, she was very uneasy about this man. If he had anything to do with sending the wolves after her, did he want her dead? If not, why then? And if he wasn't involved, why was a piece of his waistcoat snagged in a branch at the scene? Hopefully, when he and Bethany were finally wed, Marissa would have no further trouble with the man.

"So, Bethany," Marissa said. "Have you seen a seamstress about your gown?"

"Oh, yes, Majesty," she said excitedly. "It will be so beautiful."

"I'm sure it will rival anything I could hope to wear," Marissa said.

"Oh, no, Majesty," Bethany said, somewhat anxious. "I could never have anything grander than what Your Majesty would have."

Marissa chuckled at the obvious discomfiture Bethany felt at the idea of besting the queen. "It is quite all right, Bethany," she said. "For your wedding you deserve to outshine everyone in the kingdom."

"But surely not the queen herself," Bethany protested.

"All right," Marissa said. "Let's just say it will be grand. So, tell me about it."

"Yes, Majesty," she said, and proceeded to describe her wedding gown.

Forsythe was obviously not interested in the conversation, and began looking off in the distance, a completely distracted look on his face.

When Bethany finished describing her gown, Marissa asked Forsythe, "So bridegroom, what do you think of your bride's gown?"

Without looking at either woman, he replied stiffly, "Sounds lovely."

"Good," Marissa said, refusing to be baited. "Now, let's discuss the reception menu."

"I am sure anything you ladies decide will be fine," Forsythe said mechanically.

"I thought you would say that," Marissa said.

Bethany seemed disappointed, but refused to let it dampen her mood. "I think we should offer all sorts of meats," she said. "But especially quail. It is such a delicious bird."

"That is a great idea," Marissa agreed. "Of course, we will have a pig on a spit, and also a steer. But I like the idea of the quail. We will also have the court baker think up something wonderful for dessert."

Bethany bounced up and down with excitement.

"Now, about the guest list," Marissa continued. "Everyone at court will be invited."

"Oh yes," Bethany breathed. "And may we also invite the queens from the neighboring kingdoms?"

Finally, Forsythe contributed. "Do you really feel that is necessary?" he asked Bethany sourly. "After all, you do not personally know any of them. Why would they make the journey to see someone they do not even know get married?"

"But it would be wonderful to have them here again, would it not, Majesty," she said hopefully.

"That it would," Marissa agreed. "However, Forsythe does have a point. You do not personally know them. It would seem gratuitous to invite them."

Bethany's countenance fell. "I suppose you are correct," she said.

"There will be plenty of guests with just the people from court," Marissa assured her.

"Of course, Majesty," she said, with just a hint of disappointment. "I must not seem greedy."

"That's very good of you," Marissa said. "Now for the music. I assume we will have the orchestra that played at my arrival banquet."

"Yes, Majesty," Bethany said. "They are wonderful. Everyone will enjoy dancing to their music."

"Good. Now have I forgotten anything?" Marissa asked.

"I believe we have discussed everything we need to," Bethany said.

"And what about you, Forsythe?" Marissa said. "Do you have anything else to add? Is there anything special you'd like for your wedding?"

"I am sure you ladies will take care of everything that needs to be done," he said through slightly clenched teeth.

Marissa would have thought it amusing if she didn't have such suspicions about his motives.

"Well then," she said. "Since that concludes our meeting, why don't you two go have the palace cook prepare you a nice meal. Perhaps you could find a nice quiet nook in which to be alone and enjoy each other's company."

Bethany bounced up and down, and said, "That sounds wonderful, Majesty." Then she turned to Forsythe and said, "Will that not be wonderful? Let us have some wine and a nice dinner." And she tugged his arm and led him out the door.

As he walked away, Marissa heard him mutter, "It is what I live for."

FIFTY-ONE

Roland returned from his search for the wolves, and entered the great hall, striding to where Marissa sat upon her throne. He bowed at the waist.

"Welcome, Roland," Marissa said. "Tell me what happened."

He shifted from one foot to the other uncomfortably, and said, "Nothing happened."

"What do you mean?" Marissa asked. "Weren't you able to find them?"

"No, Majesty," he said. "They are elusive indeed. We encountered wolves along the way, but none were scarred as you said the lead wolf should be. We met with a few different packs, but none numbering five in total."

"So, you weren't able to find my wolves," she said, more to herself than to Roland.

"No, Majesty," Roland said in reply. "Even Bruno the huntsman could not ferret out the pack we were sent to find."

"You spent three weeks scouring the forests around Avonridge, and could not locate the pack of wolves, the pack that has, as far as we know, only attacked me."

"Forgive me, Majesty," Roland said. "Give the order and I will spend the rest of my days searching for that pack."

"No, no," she said, and waved a hand as if swatting a fly. "If the huntsman was not able to locate them, they are not going to be found. I'll have to think of something else."

Marcellus waited on the path that led to Silas' cabin. He was able to discern just enough of the younger wizard's activities to know he would be returning at any moment, and Marcellus wanted to surprise him.

Before long, he could hear someone approaching, and he cloaked himself so as not to be detected. Presently, Marcellus saw Silas walking along the path, and even though he could not be seen, he retreated into the trees to observe the younger wizard furtively. Marcellus had wanted the advantage of surprise, but he himself was surprised by the company with which the younger wizard was keeping.

Trotting along at Silas' heels, like well-behaved house pets, were five large wolves. Marcellus easily recognized the lead wolf, as it had several deep scars on his head that looked like the evidence of claw marks. Marcellus silently observed as Silas walked along, ignoring the pack of animals that had completely eluded Roland and his hunting party.

Keeping still and out of sight, Marcellus watched as Silas made his way to his cabin, opened the door and entered, leaving the five wolves standing outside. They sat down as if waiting for Silas to return. After several minutes, he did just that. Opening the door, Silas tossed out the carcass of a wild boar. Almost before it hit the ground, the scarred wolf grasped it in his jaws and dragged it off into the trees. The other wolves followed, and soon Marcellus could hear the unmistakable sounds of meat being ripped apart and devoured.

Silently, Marcellus nodded his head in understanding. Believing that the wolves would not forsake their meal, he silently made his way to Silas' door. Still cloaked, and even more angry at the unmistakable knowledge that Silas was behind the attacks on the queen and his pupil, Marcellus didn't bother to knock this time.

Throwing his hand out before him, the door to Silas' cabin burst inward as if it had been kicked. Striding in through the open doorway, Marcellus was pleased to see confusion on the younger wizard's face.

"What is going on here? Is someone there?" Silas demanded.

Remaining cloaked, Marcellus silently walked to the small table at which Silas sat. Assorted pots and pans hung on the wooden wall, and a rack with several lethal-looking knives was suspended beside them. Marcellus was glad of all the utensils; they would come in handy.

Silas rose from the table and went to close the door. However, Marcellus beat him to it, and swung his hand in a sideways motion, causing the door to slam shut. Immediately Silas stopped and turned around, searching the cabin for the source of the magic.

"Where are you, old man?" he said angrily. "You enter my home uninvited. Show yourself."

In response, Marcellus, still cloaked, raised a hand and flung it forward in a throwing motion. A large pot flew off the wall and sailed straight for Silas' head. He was able to duck at the last moment, and narrowly avoided being struck.

"I said show yourself!" Silas shouted, but fear was evident in his voice.

Suddenly, Marcellus materialized. Standing about ten feet away, he was an impressive figure. Stooped at the shoulder, but tall none-the-less, his gray beard hanging down to his waist, the gray hood of his cloak covered his head, casting his face into shadow. Dark, malevolent eyes seemed suspended within the darkness of the hood, and bore into Silas with anger and hatred.

Silas grasped in surprise and fear, but quickly recovered. He raised his hand as if to throw something, and a fireball appeared in his palm. He threw the fireball at Marcellus, who lifted a hand and swatted it away like a bug. The fireball flew to the side, landed on the floor beside Marcellus and erupted into a massive flame. Reaching to the ceiling, it began to move along the floor toward Marcellus.

Now Marcellus raised his hand, and a great wind roared through the cabin, blowing out the fire and knocking Silas back a few steps. One of the knives flew from the rack on the wall and headed straight for Silas' chest, arrow-sharp point aimed for destruction.

Silas regained his footing and spun to the side, just avoiding the knife that struck and wedged itself into the wooden door. He then clapped his hands, and the table became animated. Its four legs acquired joints and it ran at Marcellus, intent on trampling the old man.

Marcellus put up his hand in a 'stop' gesture, and the table disintegrated, breaking apart into hundreds of broken bits of wood. He then waved his hand before him as if performing a magic trick, and the hundreds of pieces of wood all became living snakes. As one huge swarm, the snakes slithered in the direction of the younger wizard, who drew back in fear.

Obviously frightened, Silas raised both hands before him, and the snakes paused. They did not, however, vanish or return to their wooden state.

Desperately, Silas yelled, "Attack!" and almost immediately, the wolves were at the door, growling and snarling, and scratching to get in. Holding the snakes at bay, Silas couldn't use his magic to let the wolves in, so he inched his way closer to the door to open it for them.

Marcellus made a pushing motion, and the snakes surged forward. Just as they were about to overrun the younger wizard, he managed to yank open the door and the lead wolf leapt in and ran straight for Marcellus. Distracted, his magic failed and the snakes all returned to wood. He threw up his arm and the lead wolf was thrown aside as if kicked. The other wolves, however, were rushing in and heading for the old man. Admitting temporary defeat, Marcellus disappeared just as two wolves leapt through the air at him, missed when he vanished, and crashed into the wall.

Reveling in his victory, Silas shouted, "I have beaten you, old man! Do not make your feeble attempts on my life again."

FIFTY-TWO

Hoping to beat Marcellus to the palace, Silas hurried out his door and down the path. He wanted to get to Queen Marissa and try to convince her she needed him as court wizard and advisor. Knowing Marcellus had originally opposed her ascending the throne, he hoped he would be able to persuade her to get rid of the old man, making her realize he was just too dangerous to keep around. When he reached the palace, he spied Jasper the page, and beckoned him.

"Boy," he called. "Come here immediately."

Jasper hurried over to Silas and said, "Yes sir, how can I help?"

"I wish to see the queen. Tell her Silas requests an audience."

"Very good, sir," Jasper said, and hurried up the palace steps and in the door.

Silas made his way up the steps, knowing he would be allowed to speak with Marissa, and so waited on the landing for Jasper to return. He waited only a moment, and the page reappeared.

"Her Majesty, Queen Marissa, says you may enter," he said, and turned to lead Silas in.

Silas followed Jasper down the corridor to the great hall, all the while mulling over what he would say to her. Not only did he want to dispose of the child wizard, but now he needed to get Marcellus out of the way as well. If Marcellus had had the chance to inform Marissa that the wolves were under his command, it would be too late. He walked calmly behind Jasper, fervently hoping that Marcellus had not yet arrived.

When they reached the great hall, Silas said, "Leave us," and Jasper hurried away.

Walking with head held high and shoulders back, the air of arrogance about him was belied by the poor clothes he wore. The rough cotton shirt and leggings bespoke the everyday hardships of the common man. Silas, however, walked as if he had the world in the palm of his hand. As he made his way toward the throne, he saw the queen's young ward sitting to her left, playing with carved animals, and it took all of his composure to keep a neutral expression upon his face. Stopping approximately fifteen feet in front of the queen, Silas bowed deeply at the waist, stood up, and said, "Majesty."

Marissa was silent for several seconds. In fact, she refrained from answering for so long that Silas' cheeks paled slightly.

Finally, she said, "So, you are Silas."

Smiling broadly, he replied, "Yes Majesty. It is I." He waited for her to say something more, but when it became evident she would not, he spoke again. "I am sure you have heard of me, Majesty. After all, I am quite the accomplished wizard."

Marissa raised her eyebrows, but said nothing.

Silas continued. "Yes, I am quite talented. And very knowledgeable about the goings on within Avonridge."

"How can that be so?" Marissa asked coldly. "Since I have never seen you at court, how can you know anything about the goings on within Avonridge?"

"Ah, yes, very good question," he said. "You see, I can see things. I have visions that come to me and provide me with information."

"Provide you with information," Marissa repeated. "How interesting."

"Yes, Majesty, quite so," he said. "And these visions are not only of the past. I can see into the future as well."

"The future?" Marissa said. "You can actually see true events before they happen?" Her voice was still cool and unyielding.

"Yes, Majesty, that is true. I saw the impending visit of King Arthur a full week before he arrived. I saw the challenge of King Sorens before your coronation, even."

"Really?" she said. "And why did you not come forward until now?"

"I have never been one for court life, Majesty," he replied. "I had never had the desire to be around so many people. The masses bore me."

"And what has changed to make you seek the masses now?"

"I have something to offer you, Majesty. I have a service to perform for you."

"And just what would that service be?" she asked.

"Why, advice, Majesty," he said as if it was quite obvious. "I wish to become your advisor."

Marissa was silent, simply looking at this man in peasant's clothing asking to be given a position of privilege at her court.

Again, Silas became uneasy with the lack of a response.

"You see, Majesty," he said. "I can help to keep you safe. I can help you make difficult decisions. Whenever you are unsure of anything, I can look into the future and advise you on the best possible choice."

"You said you can keep me safe," she said. "How so?"

"Ah, yes, Majesty, I can foresee danger and advise you how best to keep out of harm's way."

She nodded. "I see."

"As an example," he said, "were I here the night of the fire, I could have warned you early enough to have avoided it completely."

"The fire," she said. "If you could have allowed us to completely avoid it, why did you not come to court and do so?"

"As I said, Majesty, I wanted nothing to do with life at court. But now I feel my talents are needed. So much so that I am willing to put my own feelings aside and serve my queen."

"How touching," Marissa said. "Your loyalty is admirable." But her voice, while not sarcastic, was obviously skeptical.

"So, my queen, do you agree that Avonridge could benefit from my involvement?"

She didn't answer right away.

"Let me offer some advice now, Majesty," he said, "to show you my sincerity. You have a wizard who seems quite familiar with the goings on at court, do you not?"

Marissa thought to herself, now here it comes. Then she said aloud, "Yes, why do you ask?"

"I only wish for the safety and well-being of my queen, Majesty. I know Marcellus, perhaps more intimately than you. He is not to be trusted."

"Oh?" Marissa said.

"Yes, Majesty. Marcellus is a trickster. He would have you believe him your friend and ally, the whole while plotting to steal your throne. He has high ambitions, and he is not above outright deception to achieve his desires. He would use any trick at his disposal to acquire his wishes. If I were to make the decisions, I would remove him from court immediately."

Marissa hated to admit it, but she still held doubts about Marcellus in the back of her mind. She wondered if he indeed might not try to steal her throne.

"In fact, Majesty," Silas continued. "I would not be surprised if it was he who set the fire that night. He may have hoped you would perish in the flames." His voice had become cloying.

Now Marissa suspected Silas was simply trying to get on her good side. She knew Marcellus would never put Kevin in danger.

"You think he may have sent the fire?" she asked.

"I believe it entirely possible, Majesty," he replied. "He would have motive."

"But you do not know for certain?" she said.

"Ah, no, not for certain," he demurred.

"But you said you see things. You said you saw the fire. How could you have seen it and not know how it was sent?"

"Ah, well, Majesty," he stammered, but then quickly recovered. "I can see things, it is true. However, I do not get each and every detail. I see events about to happen, and I am given enough information to keep those involved safe, but alas, I am not given everything."

"I see," Marissa said, raising one eyebrow.

"So, Majesty, do you think me a worthy addition to your court?" he asked hopefully.

She didn't answer immediately. He leaned forward, smiling expectantly.

"It is something I must think about," she said. "I will send for you when I've made my decision."

He drew back in obvious disappointment, but said, "Of course, Majesty." When he turned to go, his eyes fell upon Kevin, and Marissa saw unmistakable hatred on his face.

The captured soldiers of King Sorens' army sent word that they wished to speak with Queen Marissa. She made her way to the barn to grant them audience, with Roland by her side, and leading Kevin by the hand.

"Majesty, I am not comfortable with your offer to the soldiers," Roland said.

"All right," she said. "Tell me your feelings."

"I believe they will tell you they will accept your offer, agree to become subjects of Avonridge, and then rebel against you when they have their freedom."

"That is a valid point," she agreed. "I will, however, have them swear an oath. I believe they are, for the most part, honorable, upstanding men. I believe when they make a promise, most will abide by it."

"I hope you are correct," Roland said, *but seemed* unconvinced.

Walking into the barn that served as a makeshift prison, Marissa was greeted by a soldier that was obviously the spokesmen. The body of prisoners was divided into two groups. The group to the spokesman's right was larger by far, with the group to his left consisting of only about three hundred men.

The soldier bowed at the waist, and said, "Your Majesty."

Marissa nodded, and said, "You've asked to see me. I assume you have made up your minds about my offer?"

"Yes, Majesty," he said. "We have discussed it among ourselves and made a decision."

"Very good," Marissa said. "First, what is your name, soldier?"

"I am Alex, Majesty, captain of King Sorens' army."

"Okay, Alex, tell me what you have decided."

"Well, Majesty," Alex said, "we have talked at length about your offer. We discussed how King Sorens seems to have forsaken us. We talked of the benevolent manner in which you have treated us. We have come to the conclusion that you are a fair and just queen, and we believe you would treat us in the same manner as your other subjects, should we agree to live here in Avonridge."

"That makes me feel good, Alex," Marissa said. "Should you decide to stay and become my subjects, I will most certainly treat you like my own people."

"Yes, Majesty," he said. "We have not all agreed; as you can see, those to the left are separated from those on the right."

"I can see that," Marissa said. "I hope the larger group is the one that has decided to become a part of my kingdom."

"Indeed, Majesty," Alex said. "We have found your offer to be generous and acceptable. We wish to become subjects to a sovereign that shows compassion."

"Splendid," Marissa said. "I will send for the bishop and he will administer your oath."

"Oath?" Alex echoed. "I do not understand."

"You do not expect Queen Marissa to simply accept you decision without some form of assurance, do you?" Roland said. "There would be nothing to stop you from simply fleeing back to Southcross, and making war on us the next time King Sorens attacks Avonridge."

"I will give you some time to think about this," Marissa offered. "But, make no mistake, this will be an oath before God, and you will be accountable to Him for it. Now, I will send for the bishop, and when I return with him, you who wish to stay and live in peace within Avonridge may do so after you have given your oath. If you want to cling to the hope that Sorens will negotiate for your return, do not make the oath. You can continue to live in the comfort of the prison barn, with regular meals and shelter from the cold." And Marissa and Roland left the soldiers to make their decision.

FIFTY-THREE

"Majesty, may I have a word with you?" Marcellus called as Marissa was walking down the street in the direction of the palace. He smiled at Kevin and patted him on the head affectionately.

"Of course," Marissa said.

"Majesty," he said, falling into step with her, "I have found out for certain that which we had suspected. Silas does indeed control the wolves."

Marissa looked at the old wizard. "You're sure?" she said.

"Quite," he replied. "He almost seems to keep them as pets. I witnessed him feeding them, and they came to his rescue when he summoned them."

Marissa nodded. "Yes, well, that does not come as a surprise. He stopped in to see me just a few hours ago," she said.

Marcellus stopped walking, and Marissa did as well.

"He came to see you?" he said, somewhat alarmed. "What did he say?"

"Oh, he told me how valuable an addition to my court he would be. He said he felt a duty to advise me to keep me safe."

"Keep you safe," Marcellus growled. "And I suppose he believes sending a wolf pack after you would be keeping you safe."

"We did not discuss the wolves," she said. "But he did say he thought you sent the fire in Kevin's chambers."

"Absurd!" he hissed. "What trickery!"

"Relax, Marcellus," she said. "I know you would never do anything that could put Kevin in danger."

"Indeed not, Majesty," he agreed. "I will protect him to the best of my ability."

"And I know that," she said soothingly.

"So, he has asked you to accept him into court as your advisor," Marcellus said. "What was your answer?"

"I sidestepped and told him I'd think about it," she said.

"You would not seriously consider allowing him a place in your court?" he asked, alarmed.

"Not in a million years," she replied. "But I figured it was the easiest way to get him out of my palace."

Marcellus smiled. "Yes, you used your wiles. Admirable," he said.

"But that brings us to the question of what do to about him," Marissa said. "Now that we know he sent the wolves, I cannot help but wonder why. If he wants to be a part of my court, why would he try to kill me?"

"Yes, a mystery," Marcellus said. "Why indeed. Perhaps the wolves were not meant to kill you. Perhaps he used them to make you agreeable to his advice. If he could warn you of impending danger, you would be more willing to accept him."

"Perhaps," Marissa said. "But we did not discuss the wolves. He did not make any reference to forewarning me about them."

"Then I am sure I do not know his motives," he said. "But I do know the wolves obey him."

"So, again, what do we do about him?" she said.

"I will attend to Silas," Marcellus said. "Now that I know his tricks, I will not be caught unaware again."

* * *

Marissa and the bishop walked into the prison barn, followed by Roland. Little Kevin was with Marcellus, and out of the way. When they entered the barn, Marissa noticed the smaller group of soldiers had grown. Now, instead of approximately three hundred, there were more than six hundred. Still, all in all, those that would be pledging an oath to Avonridge numbered well over a thousand.

Alex bowed before Marissa and said, "Majesty, there are those who will not offer an oath of remaining in Avonridge. At first, they would swear, and then flee to Southcross anyway. However, I was able to convince them that to break their oath would be a sin before God."

"A very wise statement, my son," said the bishop. "The men will indeed be promising before God to remain in Avonridge and become loyal subjects. Should they forsake that promise, it will be a personal affront to our most Holy Father."

"Yes, Bishop," Alex said.

"Now, shall we begin?" said the bishop.

"Yes, please do," Marissa said.

The bishop addressed the crowd of soldiers that would be becoming subjects of Avonridge, and said, "Men of Southcross, you will be taking an oath that is binding before our most Holy Father, the Lord God Almighty. You will pledge of your own free will your loyalty to Avonridge and Queen Marissa.

"You will pledge to forsake the life of soldier, spending your lives as men of peace; farming, herding and raising families. Should there be any among you who do not freely make this pledge, do not blaspheme before our God. Breaking this most sacred oath before our God will condemn you to burn in the fires of Hell for all eternity."

Marissa could feel the weight of the oath these men would be pledging, and hoped the bishop's words were enough to command true loyalty. She couldn't imagine anyone of them breaking the oath after being promised the fires of Hell.

"So, men of Southcross," the bishop continued, "if you do not truly offer this pledge, do not undertake it. You may step aside without recrimination. There will be no punishment meted out should you decide not to become subjects of Avonridge. So, make your choice now, for the benefit of your everlasting souls."

Marissa was a little disappointed to see five or six soldiers break from the group and go stand with those that had already refused to take the oath.

"Now that that is finished," the bishop said, "prepare to take your oath."

Marissa saw some of the men smile, while others seemed as solemn as the bishop himself.

"Men of Southcross," the bishop said, "this is the last time you will be referred to as such. Each of you standing before me is on the verge of making a promise that will be binding upon your everlasting souls. Do you hereby promise before our Lord God, to become loyal subjects of Avonridge and Queen Marissa, to forsake your past lives and begin anew as if reborn, to hold dear to your hearts your new sovereign and defend to the death Avonridge and all who dwell within her? Do you so swear before God and Queen?"

In unison, the soldiers all said, "I swear."

Marissa's eyes filled with tears. She felt this was another victory over King Sorens, to have more than one thousand of his soldiers voluntarily break from his realm and join Marissa's.

"You are now men of Avonridge," said the bishop.

Marissa wasn't sure what she had expected, but there were no shouts of joy, no congratulatory pats on the back, no revelry at all. The men simply turned tone another, nodding their heads as if of one mind. Well, at least they seemed to accept it without hesitation. She was glad that they had made the decision to stay. She felt her kingdom became stronger because of it.

* * *

"I do not have time for your petty problems," Silas said harshly. "Things did not go as planned with the wolves. I have other matters that are much more pressing."

"We had an agreement," Forsythe said. "Your wolves were supposed to do their job."

"I assure you, I am as disappointed as you," Silas said. "In fact, even more so. You will simply be married to someone you do not love. How tragic!" he said cruelly. "I, on the other hand, must still figure out a way to rid Avonridge of an unnecessary wizard."

"I have much more to lose than you," Forsythe said. "I wish to marry the queen, while you simply desire quarters within the palace. You could become one of the serving staff to achieve that."

"I do not simply wish for quarters within the palace," Silas sneered. "I wish to dispose of a wizard that distracts the queen from her royal duties."

"How loyal of you," Forsythe mocked. "But you have another wizard to contend with as well, do you not? Marcellus has taken the little brat into his care. He tutors the boy in the ways of magic."

Silas glared at Forsythe, but said nothing.

"So, not only do you have to get rid of the boy, but you need to overpower Marcellus to do it," Forsythe said. "Not an easy task, I think."

"I can handle Marcellus," Silas said.

"Like you handled Marissa?" Forsythe said. "You could net even convince her that Marcellus wants to steal the throne. How do you plan to convince her to get rid of the boy when you cannot even cast enough doubt to convince her to get rid of the old man? I really do not see you succeeding at anything of importance," he taunted.

"You concern yourself with your own situation," Silas said, "and leave the wizards to me. As I said, you do not even have what I would consider a problem. So, you are being forced to marry. That is little more than an annoyance. Not worth losing a night's sleep over."

"I will die before I marry anyone other than Marissa."

"How prophetic," Silas said darkly. "I see exactly that."

"Do not try to frighten me," Forsythe said. "I will sit upon the throne, it is my destiny."

"Just as your destiny was to marry Queen Gertrude?" Silas said. "Come now, Forsythe, do you truly believe you are meant to be king? Even if, by some miracle, Queen Marissa did relent and decide to marry you, you would not be king, you would simply be Royal Consort. There is no power in Royal Consort."

"Things can change," Forsythe said. "Marissa could be convinced to change the way things are done. All she needs is the right persuasion."

Silas laughed. "You truly do believe you would be king," he jeered. "You truly believe Marissa would give up her power as queen and defer to you. You are mad."

Quick as a flash, Forsythe drew a knife out from the back of his waistcoat and put it to Silas' throat. "Never call me mad!" he hissed.

Silas laughed again, waved his hand, and Forsythe flew backward across the room of the magician's cabin, slammed into the wall and crumpled to the floor.

"You think too highly of yourself yet again," Silas said.

Forsythe stood up slowly, anger blazing in his eyes. "You will not call me mad," he said.

"As you wish," Silas said lightly. "It makes no difference to me what title you give yourself."

"You just do as we agreed," Forsythe menaced. "Convince Marissa to marry me."

"Apparently that is more difficult a task than I had originally foreseen," Silas said. "She simply is not interested."

"So, use your magic to make her interested," Forsythe demanded.

"Unfortunately for you, magic does not perform in that manner. You must be the one to make her interested, and so far, you have failed." His voice trailed off into an obvious insult.

"Just as you have failed in your attempts to make her realize she needs a protector," Forsythe said. "Your wolves do as they are instructed, you said. Yet, they cannot even frighten a woman sufficiently to allow her to need comfort."

"I cannot be blamed if she is courageous enough to not need comfort. She is quite self-sufficient, and may simply not need for anyone to hold her hand like a frightened child. And that, my friend, I cannot change."

"Then you are useless," Forsythe spat, and flung open the door and left.

FIFTY-FOUR

Silently, and cloaked from sight, Marcellus made his way through the woods toward Silas' cabin. This time, however, he would not be meeting with the younger wizard. In fact, if all went as planned, Silas would not even know Marcellus was there. This visit was for the sole purpose of communing with the wolves.

Marcellus walked stealthily to the door of Silas' cabin, placed an invisible hand upon the wood, and detected the cabin was empty. Turning to look out into the trees, he materialized so the wolves would be able to see him. Taking a pinch of powder out of a leather pouch that hung around his neck, he blew the powder in the direction of the trees, and commanded, "Come."

Within the span of two heartbeats, Marcellus heard the footsteps of several animals walking through the leaves and underbrush of the forest. He waited patiently, and soon he could see several sets of eyes peering at him from within the trees.

Taking another pinch of powder from the leather pouch, he blew it in the direction of the wolves, and again commanded, "Come."

Calmly, five wolves walked out of the trees and ambled up to stand before Marcellus. He looked at the one that led the way, the largest, and saw several scars on its head and muzzle. He hadn't had any doubts that these were the wolves he sought, but was satisfied at the proof none the less.

He waved his hand in a circular motion, and said, "Sit."

Immediately, all five animals obeyed, sitting on their haunches like well-behaved family pets. Their tongues lolled out of their mouths, giving each the impression it was grinning up at Marcellus. He looked down at them and smiled.

"From this day forward," he said, "your allegiance will be to me." He took out yet another pinch of powder and blew it over the five wolves. One reached up and snapped at the powder, as if trying to taste it. He then licked his chops and again took to grinning at Marcellus.

"You are mine," Marcellus said, "and you will obey only me."

The lead wolf gave a yip as if in acknowledgement.

Suddenly, Marcellus could feel the presence of Silas. He cloaked himself just as the younger wizard appeared on the walk. Stepping back, Marcellus observed Silas walk to his cabin and grip the door handle.

"What are you doing here?" he said to the wolves. "I have no need of you at the moment. Go until you are called."

Marcellus watched in silent satisfaction as the wolves simply looked at Silas, but made no move to obey.

"I said go," Silas said forcefully. "I will call you when I have need of you." And he waved his hand to shush them away.

The wolves continued to sit, refusing to obey. Not wanting Silas to become suspicious, Marcellus waved an invisible hand at the wolf pack, and as one, they rose and jogged off into the trees.

"That's better," Silas said, and entered his cabin, closing the door behind him.

Marcellus was very pleased at the situation, and walked away from Silas' cabin smiling.

Marissa gave Ruth the daunting task of integrating the thousand plus soldiers into peaceful existence within Avonridge. To Ruth's surprise and great relief, the people of Avonridge seemed to welcome the new-comers without hesitation, making a difficult endeavor somewhat easier.

Some of the soldiers wanted to farm, and there were villages that welcomed the extra hands for planting and reaping. Some wanted to tend stock, and there were villages that needed men for herding. With Marissa's permission, Ruth allotted several heads of steer, pigs, goats and sheep to the herding villages to help offset the extra mouths to be fed.

All in all, Marissa was very pleased with the way Ruth was able to find homes for each new member of her kingdom.

The soldiers that did not choose to take the oath of allegiance to Marissa continued to exist in relative comfort, staying in the prison barn and being fed regularly. Marissa was true to her word and there were no recriminations for those that did not wish to become subjects of Avonridge.

Marissa led Kevin through the woods in the direction of Camille's wooden hut. She thought it was time the old seer finally met the young boy, as he was going to be a part of the kingdom of Avonridge for many years to come.

As she had promised Roland, Marissa was not without protection; Sir Erick accompanied them on their journey. He walked a few paces behind the queen and her ward, hand on the hilt of his sword, ready for anything that might occur. Or so he thought.

The late spring day was comfortably warm and fragrant. The path through the forest was cushioned in pine needles and moss, making it feel as if they were walking on a carpet. Small purple flowers grew beside the path here and there, and the buzzing of honeybees could be heard off in the trees.

"It is a beautiful day, isn't it, Sir Erick?" Marissa said over her shoulder.

"Yes, Majesty," he said. "A fine day for a visit."

"I am sorry to have to drag you along with us," she said, "but under the circumstances, I thought it was best. There is no telling what that damned Silas has up his sleeve now."

"My only purpose is to protect you, Majesty," he said. "Do not trouble yourself."

She nodded, feeling comfort knowing that was true.

Walking along, Marissa was fully enjoying the day. The air was fresh with the scent of new life, and birds were singing, calling to their mates. As they walked, Marissa allowed herself to be lulled into a sense of serenity, and didn't notice when the birds stopped singing. In fact, all sound ceased except for the footfalls of the three travelers.

"Majesty," Sir Erick said softly. "Do you hear that?"

"I don't hear anything," she said dreamily.

"Exactly," he said. "The sounds of the forest have halted. That is a bad omen."

Realizing he was correct, she slowed down, and then stopped altogether. A feeling of being surrounded overcame her, and a chill ran up the back of her neck. She looked around, searching for anything out of the ordinary in the trees.

"You are right," she said. "I do not like this."

Suddenly a loud crack reverberated through the air, and a large tree branch swung down and across the path. It didn't fall to the ground, but swung as if suspended by a rope. The trio was caught completely by surprise, and didn't have time to react. The branch slammed into Sir Erick, connecting with his shoulder and the side of his head, throwing him to the ground like a rag doll, and then crashing down to land upon his prone body. He lay motionless.

"Oh no!" yelled Marissa. "Sir Erick!"

But before she could help her dear friend, a vicious squeal could be heard through the trees. Marissa didn't recognize the sound immediately, but it was followed by a second squeal, then a third, and they were getting closer very quickly. The unmistakable sound of a large animal crashing through the underbrush of the forest filled the air, and now the mysterious animal began grunting.

Finally, Marissa realized what kind of animal was coming their way; a large pig, or rather a wild boar. She knew from books that boars were dangerous animals. Pigs, she knew, ate anything, even meat, so it stood to reason that a wild boar would also.

As the sound of the grunting animal drew closer, Marissa looked around for some kind of weapon. Sir Erick's sword was strapped to his waist, but the tree branch lay across it, trapping it.

"Mama!" screeched Kevin, and she turned to see the boar burst from the trees onto the path about fifty yards away, and run straight at them, grunting as it ran. Wicked looking tusks protruded from its lower jaw, and it had its lips pulled back revealing teeth that looked as if they could rip the meat off a bone with very little effort. The animal was huge, probably weighing close to a thousand pounds.

Torn between trying to free Sir Erick's sword, and needing to protect Kevin, Marissa hesitated only a moment. Grasping the little boy, she fairly threw him to the side and out of the way of the charging animal.

Kevin landed on his feet and jumped behind a tree, then turned to face the danger. With obvious fear in his eyes, he yelled, "Go away!" and thrust his hands at the beast in a punching motion. The boar had run several paces past Marissa, and was spinning toward Kevin to make another attack, but with Kevin's command, it was knocked to the ground. The boar immediately jumped to its feet, and resumed its run at the little boy.

"Go away!" Kevin shouted again, and again the boar was knocked down. However, it again came to its feet and resumed its attack.

Marissa realized the beast was ignoring her, trying only to get to her precious boy. She used the time to wrench the tree branch off Sir Erick, and grasped the hilt of his sword. It slid free of the sheath easily, but it had been fashioned for a very large, strong man, and was quite heavy. As Marissa tried to raise the sword to attack the boar, she almost couldn't lift it.

The boar was getting closer to Kevin. Each time he commanded it to go away, it would only be knocked down, not backward. Marissa would have thought it would turn its attention to her, or even the prone Black Knight, but it kept going after Kevin, almost as if it was on a mission.

That thought struck her, and anger flooded her mind. First the fire, next the wolves, and now the boar. This would come to an end, and it would end now! With the strength of a rage that burned white-hot, Marissa lifted Sir Erick's sword in both hands above her head, point aimed at the large beast that sought to harm her little angel, and charged.

The boar rose to its feet again, and leapt at Kevin just as Marissa ran at it. As she ran, she bellowed out, "No!" at the top of her voice. The boar was finally distracted, and turned to face the enraged woman that bore down on it, just as Marissa thrust the sword downward. The tip of the sword caught the boar just in front of its left shoulder, directly behind the neck, and sunk cleanly into its flesh. The force of the thrust drove the blade almost two feet in, and Marissa hoped it penetrated enough to pierce the beast's heart.

Whether she reached its heart or not, the boar stopped in its tracks. It stood for a moment on unsteady legs, squealing out heart-rending squeals, and finally dropped to the ground, silent.

Marissa leaned her head against the tree that Kevin was hiding behind, trying to catch her breath. She was bathed in sweat, and her heart was pounding in her throat.

"Mama," Kevin said softly, and she looked at the cherubic, little face.

"It's all right now," she said soothingly. "It's all right."

"I know, Mama," he said. "You fixed it." And he grinned up at her.

Overcome with the tension of the situation, and the relief of being free of it, Marissa started to laugh. Kevin ran to her, laughing as well, and she scooped him up in a big bear hug.

FIFTY-FIVE

"This will be coming to an end," Marcellus said as he stood before Marissa in the great hall. "I will not allow this danger to little Kevin to continue a moment longer."

Marissa sat on her throne, anger anew in her heart after describing their ordeal to the magician. She too, knew it was time to put a stop to the threats to Kevin's life. She wanted to call Silas to court and accuse him openly, but Marcellus argued against it.

"He will only be able to wriggle out of it, Majesty," he said. "We have no proof with which to accuse him, and he may possibly be able to get public opinion on his side. It is not worth the trouble that may arise."

"I do not give a damn about public opinion," Marissa said hotly. "I am the queen and my word is law."

"This is true," Marcellus said. "However, should there be subjects that do not believe Silas capable of such heinous acts, they may believe you are only persecuting him for some personal reasons. They would then begin to wonder if they may be next. No, it is not worth the trouble. I have said I will take care of Silas, and the time has come to act decisively. I will toy with him no longer. I will take care of him, and we will be rid of him."

Marissa was suddenly struck with guilt at the idea of being 'rid' of Silas. Was she actually condoning the idea of Marcellus killing him? Should she perhaps banish Silas from Avonridge instead? But then she thought about the fact that he had tried to kill Kevin three times. No, it was not necessary to feel guilt, the guilt lay with Silas. She was simply condemning an attempted murderer.

* * *

Marcellus walked down the path to Silas' cabin cloaked from view. When he reached the cabin, he placed his hand upon the door, and confirmed that Silas was not at home.

Turning toward the woods, Marcellus uncloaked and called, "Come." All five wolves obediently walked out of the trees and up to the old magician, tongues lolling out of their mouths, completely at ease.

Marcellus waved his hand over the wolves, whispering and chanting. Slowly, the wolves ceased their relaxed breathing and began to grow tense. Marcellus continued to chant and wave his hands, and the wolves became more agitated. They began to pace back and forth before the old man, every now and then issuing a short, soft growl. Occasionally one would nip at another, and be snapped at in return.

Marcellus continued to work the wolves up into a near frenzy, until he was satisfied. He then placed his hand on the door of the cabin, and opened it.

"Inside," he commanded, and the wolves obeyed, growling and snarling. Marcellus then cloaked himself again, and backed into the woods to wait. He didn't have to wait long.

In a few minutes, Silas came walking up to the cabin, grasped the door handle and pushed open the door. He entered the cabin and closed the door. In seconds Marcellus heard what he expected to hear; the loud, vicious snarls of wild animals attacking a victim. Silas could be heard above the commotion of the wolves, frantically yelling, "Down! Down!" And when that failed, pitiful screams of fear and pain.

It was over in moments. The screaming and snarling quieted and all became silent and still. Marcellus had no desire to see the end result. Knowing that the wolves were now dangerous to everyone, he waved his hand at the cabin, and commanded, "Sleep." He then drew a circle in the air with a finger, and small flames traveled around the wood at ground level, completely encircling the cabin.

In moments, the small flames traveled up the wooden walls, and before long, the entire cabin was engulfed. Marcellus watched for hours as the cabin blazed, consuming everything, and everyone, within. Presently, the intensity of the fire diminished, and the flames died down. Finally, the wood became a pile of red, hot embers. Marcellus had seen enough, and he turned from the scene of vengeance and death, and walked away.

Forsythe stood on the path to Silas' cabin and stared at the pile of burnt logs and ashes. The fire had obviously been out for over a day, and so there was no heat emanating from the ruined structure. He ventured a few steps closer, silently debating whether he would inspect the rubble for signs of the magician's body, but he knew he really didn't want to see anything of that sort.

He stood for several long moments. He had come to tell Silas that his latest attempt at getting rid of the queen's ward had failed yet again, but seeing the burned cabin, realized he was now alone in his endeavor. Kevin was still safe within the palace walls, and Marissa still had no need for comforting. Forsythe would have to rethink his plans.

He turned away from the wreckage of the cabin and walked back to court. There was the possibility that Silas hadn't perished in the fire. It was possible the fire had started by accident while the magician was out, but Forsythe wouldn't put too much faith in that idea.

He felt sure the fire had been set, and realized the only one who would do so was Marcellus. Knowing the old wizard had taken such drastic measures frightened Forsythe. If Marcellus suspected Forsythe was in league with the magician, Forsythe could be in just as much danger. If Marcellus knew Forsythe had been in on the plan to eliminate the child, Marcellus could take the same vengeance upon Forsythe.

Forsythe didn't know if he should alert Marissa to the circumstances at Silas' cabin. If she knew he had been to visit the magician, she might realize they were cohorts. He wondered if Marcellus would advise the queen exactly what had transpired. Would the old magician make the queen fully aware of just how vicious he had been?

Forsythe couldn't be sure just what Marissa knew, and what Marissa may have ordered. She herself could have ordered the assassination of the magician. Forsythe believed she would never take so drastic a step, but if she had found out Silas was the one behind the attacks, she may have become angry enough to do just that.

Now Forsythe didn't know if he should go to Marissa and try to discern exactly what she knew, or stay out of her way for fear of tipping his hand. If she suspected he was involved, she could turn her anger toward him. No, that was not a possibility he wished to consider. He should probably give the queen a wide berth for the time being. Until he could figure out his next move, he would stay out of the way, and away from Marissa's attention.

Marcellus did not go to Queen Marissa to advise her of the deed. He didn't want her to be burdened by the knowledge, and perhaps feel guilty. There was nothing to be done about it, it was accomplished, and so there was no need to cause her any more unnecessary anxiety. He would allow her to continue to keep her watchful eye on the child, and in a few days' time, simply mention that the threat was past.

He made his way to the palace to engage little Kevin in his lessons. The boy was showing amazing abilities. Not only could he cloak himself and walk among the palace fold undetected, but on one or two occasions he even cloaked himself so completely that Marcellus had lost sight of him. The only way the old wizard had been able to locate the little tyke was by sheer instinct.

Kevin's magical talent was progressing at an amazing rate, and Marcellus was gratified each time the young apprentice mastered a new feat. Marcellus did not like letting the boy exert influence over others yet. Still a child, Kevin was much too young to fully understand the gravity of such power, and even though he could get the ladies to take a 'nap' every now and then, Marcellus tried to keep Kevin's mind on inanimate objects, and perhaps an animal or two.

For the most part, however, Kevin seemed satisfied with controlling things, and didn't seem to feel the need to influence people beyond the simple suggestion of going to sleep. Since his lessons with the ladies-in-waiting had been temporarily halted, he was spending more time with the old wizard, and fully enjoyed the opportunity to perform his magic tricks.

Not yet having told Marissa that the boy was safe now from danger, Marcellus kept up the pretense for a few days. Soon, he would make Marissa aware that there was no more need for worry, and let her know the watchful eye could be eased somewhat. Unfortunately, Kevin was proving to be a spirited child, and would need a strong hand in any case. But at least she would no longer need to worry for his life. That should be most welcome news indeed.

FIFTY-SIX

Marissa and Bethany were busy planning the wedding that would be taking place in three weeks' time. Bethany was in a constant state of joyful anticipation, and Marissa was bouncing back and forth between happiness for Bethany, relief that Forsythe should soon be out of her hair, and trepidation that he might yet turn his anger toward the young woman.

She was also still mulling over the idea that he could be working with the magician Silas. Having no proof, only suspicions she couldn't act on, she wondered if she should simply come right out and ask him. Perhaps she would catch him off guard sufficiently to cause him to slip and admit something, anything. But she quickly brushed that idea aside, knowing Forsythe was too cool a character to be tricked that easily.

Besides, she couldn't figure out exactly why they might be working together. Silas seemed to want Kevin dead. Forsythe couldn't want Marissa dead if he wanted to marry her and steal her crown. If Marissa happened to get killed, she was sure Marcellus would manage somehow to ascend the throne. Forsythe had absolutely no claim to it, and so couldn't actually want her dead, could he? No, she was sure he didn't want to get rid of her, and so was still frustrated in trying to figure out his motives.

She was relieved, however, to not be bothered by him during the wedding preparations. He was making himself scarce, and she figured it was to stay as far away from his bride-to-be as possible. Unfortunately, with him out of sight, she couldn't be sure if he was up to something else. She wished she could keep an eye on him, but still keep him out of her hair.

Not wanting to dampen Bethany's happiness, Marissa kept her suspicions to herself. She and Marcellus could discuss the matter in private later. Now was a time of celebration, and so she helped Bethany with the invitations and decorating details. All in all, it was a festive time, and Marissa enjoyed helping out.

With the wedding of Bethany and Forsythe fast approaching, being one week away, Marissa was beginning to tire of all the last-minute details being thrust upon her, and finally told Bethany she had a free rein to make whatever decisions she deemed necessary. Being so close to the actual day, Marissa figured there couldn't be too much danger. Besides, Ruth was still around to keep things running smoothly.

Walking through Avonridge Proper with little Kevin strolling at her side, she was enjoying the warm spring air, when Marcellus hobbled up to her.

"Majesty, a word please," he said.

"Of course," she said.

"I just wanted to discuss with you the fact that there have been no further incidents of unexplained danger," he said. "It has been weeks since the last occurrence, and I want you to know there will be no others."

"No others," she echoed. "How can you be so sure?"

"It is decided," he said. "Things have come to pass."

"Things. What things?" she demanded.

"Do not trouble yourself, Majesty," he said. "Some things need not be under the scrutiny of the royal eye. Some things should be allowed to simply transpire."

"I don't like that, Marcellus," she said, darkly. "Did you do something I should know about?"

Instead of answering the question, he said, "If I may, Majesty. I have pressing matters to attend to. I bid you farewell for now." And he turned and hobbled away.

Marissa watched him go with a cold feeling in her stomach. Had he done what she feared? Had Marcellus killed Silas? She didn't know what to do. Should she demand to be told, or let it go? As queen, she could put traitors to death. And having made several attempts on Kevin's life absolutely made Silas a traitor in Marissa's eyes. But he hadn't been given a trial. He hadn't even been formally charged with a crime. And Marissa hadn't ordered Silas put to death. Did that make Marcellus a murderer? He was only acting in defense of little Kevin.

She decided to treat it as the elimination of a traitor. And since she wasn't positive Silas was indeed dead, she decided to wait until someone declared he was, to even make a formal ruling on it.

Not being able to quell Kevin's abilities any longer, Marissa knew it was time to make them known to the ladies-in-waiting. Since they were schooling him on earthly subjects, they spent a good amount of time with him, and on more than one occasion he had exhibited some form of magic that made the ladies question Marissa. So, one morning, she decided explain things to them.

Gathered in her royal chamber's sitting room, the women all stood about talking and gossiping, sniping about that lord's indiscretion with a tavern wench, or that lady's horrendous taste in garish attire. When Marissa walked in, she actually had to tell them to quiet down.

"Ladies, please," she said, holding up a hand for attention. They quieted down, but not as quickly as Marissa thought they should. She would have to speak to them about protocol.

"Ladies," she said again. "I have been wanting to make you aware of a situation." She walked to the lounge set against the far wall and sat down. The ladies all gathered around expectantly.

"I have seen the way you have all accepted your duties of teaching Kevin," she began honestly. "And I want you to know I appreciate it. I understand you were hesitant at first, due to the manner in which he came to live with us, but you have risen to the task admirably."

Murmurs of agreement were heard around the room.

"And since you have accepted him so nicely," Marissa continued, "I believe there is something you need to know." She paused, and the ladies all leaned forward in anticipation.

"As some of you may know," she said, "Kevin has been spending time with Marcellus."

"Yes, I have noticed, Majesty," Jane, the older lady said. "I have been meaning to speak to you about that. I do not believe it should continue."

"I understand how you feel, Jane," Marissa said gently. "However, there are circumstances involved that you ladies need to be aware of. The reason Kevin has been spending time with Marcellus, is because Marcellus is teaching Kevin, much the same way you ladies are."

"What could Marcellus possibly teach Kevin?" Jane asked. "Does he know anything other than causing mischief with his magic?"

"That is a very good question," Marissa said, smiling. "I do not know what else Marcellus could teach Kevin but magic." She let that sink in for a few moments, and heard gasps of understanding go around the room.

"You do not mean that Marcellus is teaching Kevin magic, Majesty," Louisa said, aghast.

"That could not be what she means," Jane said to Louisa. "Is that correct, Majesty?"

"That is exactly what I mean," Marissa said. "Marcellus is teaching Kevin magic."

More gasps, and a few of the ladies took a step backward, putting their hands over their mouths in surprise, or fear.

"Magic is not an acceptable lesson for a royal ward of the court, Majesty," Jane said, scolding. "You simply must rethink the wisdom behind that irrational decision."

More murmurs of agreement.

"Ladies, listen please," Marissa said, holding up a hand for quiet. "The circumstances I mentioned earlier are; Kevin has magical abilities."

Again, the ladies gasped.

"Hasn't any one of you noticed things?" Marissa asked. "Have you not been conducting Kevin's lessons and seen something that you could not explain? Hasn't Kevin disappeared from time to time?"

"Well, yes," Jane said hesitantly. "In fact, just yesterday I turned my back for only a moment and he was gone. I ran out into the hall, and he was quite far down the corridor. I decided he simply was a very quick child."

"Has he ever levitated anything?" Marissa asked, an artful gleam in her eye.

"I saw him do that," Louisa whispered.

"You did?" Danielle said. "When?"

"Oh, several times," Louisa said. "I would turn around and see his quill hanging in midair."

"Why did you never say anything?" Danielle asked.

"He would grab it quickly," Louisa explained. "I just thought I was not seeing things correctly. And, if I had told anyone of you, you would have declared me feeble-minded," she insisted.

"Well, you are feeble-minded," Danielle said, "but not because you saw things."

"Oh, be quiet," Louisa said angrily.

"Be quiet all of you," Marissa commanded. "So, we have determined that Kevin has, shall we say, abilities. The reason I am telling you this is so you will not be frightened when he uses them. It is all very natural, and he is progressing quite nicely."

"Magic," Amber breathed. "Majesty, I do not wish to be near the child if he is magical."

"Oh, grow up, child," Jane scolded. "He is the same little boy you have been teaching all along."

"Thank you, Jane," Marissa said. "That is exactly what I was going to say." Then to the ladies, she said, "there will be no change in his schedule. You will continue to teach him, but you will allow him his freedom. Within reason, that is. You must still keep a close watch on him, as he is still a child."

Amber seemed to understand, and made no further protest.

"I have no fear of the child," Jane said. "However, as I said, magic is not an acceptable path for a royal ward, Majesty. He should be discouraged in this pursuit. It is unseemly, and what would the lords and ladies of the realm think?"

"I really don't care what the lords and ladies of the realm think," Marissa said. "Kevin has been given a talent, and I fully intend to allow him to develop it. He is an intelligent child, he will grow to be an intelligent man. He has a good heart, he will wield his magic well."

"As you wish, Majesty," Jane said with a sniff. It was quite obvious she was completely against the royal ward becoming a magician, but Marissa didn't care.

FIFTY-SEVEN

There was a pre-wedding celebration the day before the nuptials, held within the banquet hall of the castle. All the lords and ladies of the kingdom were in attendance, and there was music and dancing and food and laughter. The couples twirling around the dance floor made Marissa think of old Hollywood movies depicting days gone by of kings and queens and festive parties.

Marissa wished she had someone to dance with, and understood Gertrude's decision to let Forsythe hang around. If he was the only one who would dance with the queen, it made sense to keep him within her circle. Unfortunately for Marissa, she could no longer allow Forsythe that closeness, and so was relegated to the position of royal spectator. Oh well, she could at least enjoy the fact that her subjects were having a wonderful time.

Bethany was buzzing about, chatting with this lady and that lady, fully enjoying the attention. Forsythe, however, was off to the side. Men would walk up to him and clap him on the back in congratulations. He would acknowledge their good wishes, say a few words, and the men would move along. It was obvious Forsythe didn't want to hold a conversation with anyone. What made Marissa uneasy, however, was the fact that he had an unusually smug look on his face. He looked as if he knew something that no one else knew, and that was worrisome.

The evening wore on and Marissa grew tired. All the members of court had stopped by and given their obligatory greeting to the queen, but none stayed too long to chat. They were obviously much too involved in the festivities of the party to be bothered to entertain

Marissa; they wanted to be entertained. And they were. The music was good, the food was excellent and the wine was flowing like water.

Marissa was trying to figure out an acceptable way to excuse herself without hurting Bethany's feelings, when Jasper the page burst into the great hall.

"Majesty!" he shouted in a frenzy as he ran to the throne. "Majesty! We are under attack!"

An icy cold fist grasped Marissa's heart and squeezed.

"Majesty," Jasper cried. "Soldiers are attacking!" He was wringing his hands, on the verge of tears.

Some of the ladies were standing with Bethany, and one screeched, "Oh no, Majesty save us!"

The room erupted into pandemonium. Women ran around bumping into each other as they searched for their husbands for protection. The men were equally agitated, searching for their wives. Everyone was shouting as they shoved people out of their way. Women began to weep, crying, "Oh no! Oh no!"

Marissa rose from the throne and spied Forsythe. The icy fist in her chest now reached down and grasped her stomach, knotting into a frozen ball of fear and anger. He hadn't moved from where he was standing. He was leaning against the wall with arms folded across his chest. The smug look he had been wearing changed to include a look of triumph.

"Majesty," Jasper shouted above the din of the room, "Roland requests your orders."

She looked at the page and focused on the situation.

"Has he raised the drawbridge?" she asked.

"He was doing so as I left him. Some soldiers managed to enter anyway."

She thought about running to her chambers to don her armor, but knew it would take too long.

Ruth was at her side, tears in her eyes.

Marissa turned to her and said, "Go to my chambers and bring the Dragon Slayer to me. I will be at the front doors of the palace."

Ruth spun and dashed out of the banquet hall.

Marissa ran to the front entrance of the palace, dreading what she would find there. Sir Erick appeared out of nowhere and ran beside her.

"Majesty," he called to her as they ran. "It is Sorens, come to have his vengeance upon us."

"Sorens?" she said. "Yes, I was expecting him. Damn me for letting my guard down enough for him to surprise me."

She reached the doors of the palace and stopped. In the courtyard below the palace steps was a scene to rival any sword and sorcery movie. The moon was full and shone down on soldiers in leather as they battled with swords and battle axes. Townspeople were running about, trying to get out of the way of the combatants. Swords arced through the air to crash down on shields, battle axes sliced across bodies trying to cut off limbs. Upon the parapets archers rained down arrows on the enemy outside the castle walls.

Ruth ran up behind Marissa. "Majesty, the Dragon Slayer," she said, and handed the sword to the queen. Marissa strapped it to her waist as Roland ran up the steps to her.

"Majesty," he said. "Your orders."

Marissa looked around wildly, and fought to rein in her rage. "Roland, where is the High King Arthur? Is he visiting a kingdom nearby?"

"I believe so, Majesty," he said. "His army is two days march from here."

"Send your fastest rider to alert him and ask for his help," she said. "Two days, we could all be dead in two days," she muttered.

Roland ran down the steps to obey Marissa's order.

Sir Erick said, "Majesty, you should don your armor. Your sword may be needed."

She nodded absently, unable to tear her eyes away from the battle scene in her courtyard. She slowly began to walk down the steps. She looked off to the drawbridge and saw a fierce battle for control of the chains. If the enemy gained control of the drawbridge chains, Avonridge would be lost. She ran to help her soldiers defend the drawbridge, drawing the Dragon Slayer as she ran.

"Majesty!" Sir Erick called, running after her. "I insist you retrieve your armor."

"No time," she said, and slashed at an enemy soldier as he tried to break past the defenders and release the drawbridge. The enemy soldier turned to fight the attacking queen, bringing his sword down and across in attempt to slice her in half.

She deflected his sword, and brought her own up, catching him just below the arm in the crease of his armor, causing a deep gash in the side of his chest. Blood gushed, and the soldier fell to the ground.

Marissa spun to fight another enemy soldier, but the skirt of her gown bunched up and she lost her balance. She saw the soldier's sword coming straight at her head. She tried to bring the Dragon Slayer up but realized she was not quick enough. Just when she was sure she would be cleaved in two, Sir Erick's sword came up and deflected the attacker's, driving the would-be assassin back and against the fortress wall. The Black Knight then thrust his sword at the soldier, running him through.

"Majesty," he shouted above the roar of battle. "I must insist you retrieve your armor. You are unprotected. I will hold the drawbridge."

She heard the wisdom in his words, and nodded, running for the palace steps. She hated to leave the battle and her men, but knew the battle would be over if she got killed. She ran down the corridors to her chambers, and was relieved to see Ruth already there, the suit of armor laid out on the bed waiting for her. She nearly tore off her dress, and jumped into the armor, strapped her sword back on, and ran back to defend her kingdom.

Arriving back at the courtyard, Marissa searched frantically for King Sorens. If she could find him and defeat him, she could put a stop to this assault, but she couldn't locate him. It occurred to her that he might be cowardly enough to not lead the attack, but hang back until he thought Marissa's army would be subdued enough for him to walk in and claim victory.

The archers were still defending from the parapets, and Marissa could see siege ropes with great hooks being thrown up onto the stone walls to allow the enemy to bring ladders up and overrun the castle fortress. Men with axes were cutting the ropes as quickly as they could, but more hooks were appearing at an alarming rate. Marissa sought out Roland.

Upon spying his queen, the captain-of-the-guard ran up the palace steps.

"Majesty," he said. "Your plans for defense are being implemented."

Marissa had thought about castle defenses since Sorens' first refusal of negotiations, calling upon every memory of every movie of the Middle Ages she could remember.

"Cauldrons of boiling oil," she said. "And rocks. Where are the catapults?"

"They are enroute, Majesty," Roland said. "I will make the arrangements," and he ran down the steps.

The number of enemy soldiers within the castle fortress was fast dwindling, as the drawbridge had been raised before a large number could enter, and so the battle in the courtyard was close to being over. Unfortunately, some of the siege lines had accomplished their intent, and enemy soldiers began climbing over the parapet walls to do battle with the defenders. Soldiers of Avonridge ran up to the parapets to aid their comrades, focusing their attention on defending the archers.

Soon Marissa saw great cauldrons being wheeled up to the parapet walls. Once positioned to be poured over the edge, the liquid within was lighted. The cauldrons were then tipped and the blazing liquid poured down the outside of the fortress wall. Marissa could hear men scream, and found some satisfaction in that.

Great catapults were set up in the courtyard and massive rocks were rolled in. The rocks were placed in the catapults, the arms drawn back and then sent forward, and the rocks were let go. They flew over the walls, and Marissa could feel the ground shake when they landed. She hoped each rock took out a dozen enemy soldiers.

The main focus of the battle had now moved to the parapets, as more enemy soldiers were climbing over the walls. Young boys ran back and forth, bringing fresh supplies of arrows to the archers. Soldier battled soldier, and more cauldrons were brought. Flaming liquid was poured out again, and again screams could be heard. Still, Marissa wondered where Sorens was.

Marissa thought about King Arthur. There was no guarantee he would come to her aid, and even if he did, that would take at least two days. The battle would be over by then, and decided. Would it be decided in her favor? She was not sure.

She looked for the Black Knight, and spotted Sir Erick still defending the drawbridge. Now, however, he fought three enemy soldiers, and she ran to help him. Coming up behind the enemy soldiers, she swung the Dragon Slayer, and very nearly decapitated one of the attackers. He fell to the ground, and one of his comrades turned to face Marissa. She brought her sword up and engaged the enemy for only a few moments before he too went down with a mortal wound.

Thinking this portion of the battle was over, Marissa groaned as two more enemy soldiers ran at her and the Black Knight. They fought off those two, and three more attacked. Realizing the fortress walls had been breached, she fought with a renewed anger, cursing the king that would steal her throne.

Every soldier of Avonridge was fighting in her defense. Unfortunately, they were outnumbered. As Marissa and the Black Knight fought to defend the drawbridge, enemy soldiers began to pour over the walls. The cauldrons of flaming defense were stopped, the soldiers were overrun. They fought valiantly, but were overpowered.

Several enemy soldiers ran up to Marissa and Sir Erick, and while the defenders were fighting, some managed to get behind them. Fear gripped Marissa's heart as she heard the chains of the drawbridge begin to clank as they lowered the massive gate. It would be over soon. Not even a year had passed since she came to Avonridge, and she had failed. This peaceful, prosperous kingdom had existed for hundreds, if not thousands, of years, and she was responsible for its demise.

Marissa and the Black Knight were surrounded by a dozen enemy soldiers. Around them the battle still raged, but she knew it would be over very shortly. The drawbridge was now completely down, and the enemy soldiers simply flowed into the fortress. Someone had located the imprisoned soldiers and released them. Grabbing pitchforks and shovels, they joined in the fight. The battle was done.

Marissa and Sir Erick lowered their swords. They knew it was futile to continue fighting. A funereal pall settled in the pit of Marissa's stomach. Avonridge, this beautiful, peaceful kingdom, had been captured by an invader. Sorens had promised to take Avonridge by force, and he had succeeded.

She thought back to King Arthur's visit. Merlin the Magician had assured her she didn't have to worry about Sorens, that he would not live to cause her more aggravation. How could Merlin have been so wrong? He had said that Sorens would decide his own death, but here he was now striding through the crowd of soldiers, walking straight toward Marissa. Funny, his sword was still sheathed, as though he hadn't even wielded it.

"So, Marissa," he said, a haughty lilt to his voice. "We meet again."

Sir Erick stepped in front of Marissa and raised his sword.

She put her hand on his arm, and said, "It's all right, Sir Erick. It is over."

"How true," said Sorens, his haughty voice harsh and mocking. "It is over, and you have been defeated, just as I said you would be. You should have taken my original offer. No lives would have been lost. Instead, good men have died here today, and the outcome is still the same."

Sir Erick finally lowered his sword, but remained protectively in front of Marissa.

"Now, Marissa," Sorens said, "you will bow before me and hand over your kingdom."

"I will not," Marissa said. "As long as there is breath within me, I will not willingly give you my kingdom."

"So be it!" the old king shouted. "Your subjects will witness your execution. There will be no question as to who is ruler here."

"A moment please," a voice off to the side called. Marissa sucked in her breath as Forsythe came strolling through the crowd. Yes, she thought, that is a good word; strolling. Or sauntering. Perhaps swaggering even.

"King Sorens," Forsythe said. "We have an arrangement."

Now Sir Erick sucked in his breath. "You traitor!" he spat. "I will kill you myself!" And he raised his sword and stepped toward Forsythe. Immediately four swords came up to meet him, and Marissa again grasped his arm.

"Not now," she said softly. "The time will come."

"You still have high hopes," Sorens said. "There is nothing to be done. You are conquered. Accept it and you will live. Fight it and you will die." And roughly reached out a hand and took the Dragon Slayer from her. Another soldier disarmed the Black Knight.

"Now, about our agreement," Forsythe said. "We agreed I would wear the crown, in exchange for telling you of the best time to attack. I advised you when the queen would have her guard down because of the party."

"You wear the crown?" Marissa said incredulously. "Have you lost your mind?"

Forsythe smiled at her. "Dear lady, you should have agreed to my generous offer of marriage. You would still be queen, even though I would be king. Now, I will be king, and you will be nothing."

The Black Knight growled behind his mask.

"Yes, you will be king," Sorens said, "but subjugated to me. You will sit upon the throne and offer recompense to me."

"You couldn't steal the crown by getting me to marry you," Marissa said, "so you made a deal with the devil."

"As I said, Marissa," Sorens said, "acquiesce and hand over your kingdom willingly, or face execution."

"And as I said," she countered, "I will not do so willingly."

"You have made your decision," Sorens said. "Tomorrow at daybreak you will be brought before the entire kingdom, and beheaded."

Marissa was silent, the weight of the situation finally pulling her down. King Arthur would never arrive in time. She had to think of something, or she would be dead by morning.

"Now, it is late," Sorens said. "I will post guards outside your chamber doors, where you will retire for the evening."

Soldiers moved to escort Marissa to the palace, and Sorens turned to the crowd.

"Soldiers of Avonridge," he called. "Your queen is going to be handing over the kingdom one way or another. If she has a change of heart during the night, she will live. If she persists in denying me my right as victor, she will die."

Murmurs went around the crowd as the soldiers of Avonridge didn't know exactly what to do.

As Sorens' soldiers began marching Marissa toward the palace, Sir Erick fell into step beside her.

"Oh no, Sir Knight," Sorens said. "Your duty as protector is done."

"As long as I live," he said, "it is never done. I will accompany my queen."

"No, you will not," Sorens stated. "You will remain here." And several more soldiers stepped between the Black Knight and the queen.

Marissa turned to him, and said, "It's all right, Sir Erick. I will be fine."

He was obviously uncomfortable, but could do nothing.

Sorens walked out to the center of the courtyard, preparing to address the crowd. Forsythe walked with him, a step behind. Suddenly, a black shadow fell across the courtyard, blotting out the moonlight. Soldiers of both armies looked up and cried out in fear. Balvindor was flying just above the courtyard, swooping back and forth.

"The dragon!" someone screamed. "Majesty, protect us!" someone else yelled.

Balvindor landed on the stones of the courtyard, and Marissa turned to face the beast with no weapon or shield, but Sorens stepped up.

"I will protect my kingdom from this menace!" he bellowed, raising the Dragon Slayer. He ran at Balvindor in attack, swinging the sword back and forth. Like a flash of lightning, the great beast's tail whipped out and caught Sorens across the throat, decapitating the old king, and his lifeless, headless body fell to the courtyard stones.

The soldiers began to panic, running back and forth or around in circles, trying in vain to find someplace to hide from the awful beast that had just killed the conquering king. Sir Erick took advantage of the pandemonium and located his sword, yanking it from the hands that had taken it from him. Panicked cries of "The king is dead," quickly ran through the crowd, and the panic spread.

Marissa ran to the fallen king and grasped the Dragon Slayer. Turning to face the dragon, she said, "Enough, Balvindor. You have done enough here today."

"Indeed, dear lady," the dragon replied. "My work here is done," and he leapt into the air and flew off into the night sky.

"Soldiers!" Marissa called. "The danger is past. It is time to stop and calm yourselves."

Some of the soldiers became less agitated, realizing the circumstances had drastically changed. They could see the body of the dead king, and knew the tide had turned, and now were uncertain what to do.

"Soldiers," Marissa said again. "Your king is dead; you no longer have a sovereign. You are within the kingdom of Avonridge, and therefore under my rule."

The soldiers seemed to realize she was correct, and all hostility between the two camps ceased.

"Since Sorens has no heir," Marissa said, "you will be allowed to return to your kingdom and serve your queen until a successor has been chosen for you."

Just then Forsythe rushed up. "You are not the ruler here anymore, Marissa," he said. "You have been defeated and I am now king!"

"I was defeated," she said, angrily swinging the Dragon Slayer to point at his throat, "but not by you! If you think to take my crown and my kingdom, do so yourself."

He paled visibly, and backed up, realizing the danger he was in.

"You have proven yourself to be a traitor," Marissa said. "You will stand trial, and when you are found guilty, you will be condemned to death."

Forsythe shook his head. "No, I am king now," he insisted weakly. "Sorens defeated you."

"Our king agreed to put him on the throne," someone from the crowd said. "He should be king of Avonridge."

"He is a traitor," shouted another. "He plotted against his queen. He should be put to death."

"As I said," Marissa said, "if he wants to steal my crown, he can do so himself. Someone give him a sword." And she held the Dragon Slayer up in challenge.

A soldier tried to hand Forsythe a sword, but he refused to take it, throwing his hands up as if burned by the metal.

"I do not have to fight for it," he said, but the sweat on his face was visible even in the moonlight. "Sorens defeated you, so I am king of Avonridge."

"Sorens has not defeated me," Marissa challenged, "he merely disarmed me temporarily. And he is dead. So, I retain my kingdom."

Forsythe stepped back, obviously frightened by this turn of events. He started to walk away, but Marissa called him back.

"Forsythe, you are not going anywhere. You will stand trial for your actions."

"I will do no such thing." He said, as he made his way through the crowd.

Marissa noticed movement up on the parapet wall, and an arrow sailed through the air, striking Forsythe in the throat. He fell to the ground and lay motionless. Marissa looked up to try to see who had sent the arrow, but could not make out the culprit.

"Traitor!" shouted someone on the parapet, but still she could not see who it was.

She knew she must do something to settle the situation. "Soldiers of Southcross," she called. "Your king is dead. As I said, you can return to your kingdom."

"But we defeated you," someone called uncertainly from the crowd. "Your kingdom now belongs to Southcross."

"Yes, we should claim it for our queen," someone else said.

Sir Erick stepped up and shouted, "Should you try to claim this kingdom, Queen Marissa will summon the dragon."

Murmurs went around the crowd of, "no, do not summon the dragon," and "not the dragon."

Satisfied his threat had worked, the Black Knight said to Marissa, "Roland and I will see to the disbursement of the soldiers."

"Yes," she said. "The soldiers that were captured during our previous battle are free to go as well. I do not have to worry about Southcross for now."

"Without an heir to the throne," Sir Erick said, "Queen Prudence will rule for a while. She should not wage war on Avonridge, she is a peace-loving woman."

"Good," Marissa said. "But we'll fortify our defenses just the same. I will not be caught off guard again. I failed here, and if it wasn't for Balvindor coming to my rescue, Avonridge would be lost. I will not let that happen again."

"Very good, Majesty," Sir Erick said.

Roland walked up to Marissa and said, "Majesty, King Sorens' men are panicking. They know not what to do."

"Send word the battle is over," she instructed. "They are free to return to their kingdom."

"Yes, Majesty," he said, and began walking through the crowd, saying, "Go back to your home, soldiers, you are no longer needed here."

Slowly, quietly, the crowd dispersed. Marissa watched the soldiers of Southcross walk away. Avonridge's soldiers all waited to insure there would be no more threat to their queen. Some began lifting the bodies of their fallen comrades to remove them from the courtyard and bring them to their homes for burial. Marissa was glad the dead were being tended to. She then turned and walked toward the palace, with Sir Erick at her heel.

Marissa now had the very unpleasant task of informing Bethany there would be no wedding after all.

Marissa sat astride Selene, her white mare, as she made her way toward Balvindor's cave. She held a large bouquet of fresh cut flowers. She wanted to thank the great beast, and so went alone. She wore her armor and carried the Dragon Slayer, but truly believed neither would be needed. Roland, Sir Erick and Ruth were all completely against her going alone, but Marissa insisted, explaining that she didn't want the dragon to interpret several riders as an aggressive party.

She stepped out of the trees onto the large plateau in front of Balvindor's cave. Selene was calm, and so Marissa felt certain the dragon wasn't within. She rode straight toward the dragon's lair, and tossed the bouquet onto the ground at the mouth of the cave. Satisfied, she then turned her mare and rode for home.